Marketing Analytics and Data Science

AF737293

Xiaojing Dong

Marketing Analytics and Data Science

Tools and Models

palgrave
macmillan

Xiaojing Dong
Leavey School of Business
Santa Clara University
Santa Clara, CA, USA

ISBN 978-3-032-11129-6   ISBN 978-3-032-11130-2 (eBook)
https://doi.org/10.1007/978-3-032-11130-2

© The Editor(s) (if applicable) and The Author(s), under exclusive license to Springer Nature Switzerland AG 2026
This work is subject to copyright. All rights are solely and exclusively licensed by the Publisher, whether the whole or part of the material is concerned, specifically the rights of reprinting, reuse of illustrations, recitation, broadcasting, reproduction on microfilms or in any other physical way, and transmission or information storage and retrieval, electronic adaptation, computer software, or by similar or dissimilar methodology now known or hereafter developed.
The use of general descriptive names, registered names, trademarks, service marks, etc. in this publication does not imply, even in the absence of a specific statement, that such names are exempt from the relevant protective laws and regulations and therefore free for general use.
The publisher, the authors and the editors are safe to assume that the advice and information in this book are believed to be true and accurate at the date of publication. Neither the publisher nor the authors or the editors give a warranty, expressed or implied, with respect to the material contained herein or for any errors or omissions that may have been made. The publisher remains neutral with regard to jurisdictional claims in published maps and institutional affiliations.

Cover illustration: eStudioCalamar

This Palgrave Macmillan imprint is published by the registered company Springer Nature Switzerland AG.
The registered company address is: Gewerbestrasse 11, 6330 Cham, Switzerland

If disposing of this product, please recycle the paper.

Acknowledgements

I want to sincerely thank all the people who helped make this book possible. The material in this book originated from a course I began teaching in 2011. As one of the pioneering courses to incorporate machine learning and data science tools to address business and marketing challenges, it benefited greatly from Dr. Fred Feinberg's generous advice and the guidance of my dissertation advisors, Dr. Pradeep Chintagunta, Dr. Puneet Manchanda, and Dr. Frank S. Koppelman.

I am deeply grateful to all my students. Over the past fifteen years, their questions, feedback, and curiosity have consistently shaped how I explain concepts and select examples.

I am also grateful to my colleagues at Santa Clara University for our many informal conversations over coffee and lunch, which have inspired me to refine my teaching and pursue continuous growth each year.

I also thank the reviewers and early readers whose valuable suggestions enhanced the manuscript's clarity and accuracy.

Finally, I am profoundly grateful to my parents, ZS Dong and YH Li, both retired professors who inspired me to pursue an academic career and emphasized the lasting influence a textbook can have. I also sincerely thank my husband, Jiang Zhang, and my two wonderful daughters, R. Zhang and S. Zhang, for their patience, unconditional love and support, and their informal proofreading and editing during the many months of writing and revisions.

Any remaining errors are my own.

Competing Interests The author has no competing interests to declare that are relevant to the content of this manuscript.

Contents

List of Figures

CHAPTER 1

Getting Ready for Data Analytics with R and Python

Learning Objectives

1. Understand the fundamental concepts of coding in R or Python.
2. Install and configure a programming environment on your personal computer.
3. Apply basic functions in R and Python to conduct data analysis.
4. Utilize online resources and tools to troubleshoot coding questions.

This textbook introduces multiple tools needed for solving a wide selection of marketing problems. To facilitate the learning process, both R and Python codes are provided to students for learning and homework assignments. The R scripts and their outputs are provided inline in the textbook; the data for these codes, together with the Python code and outputs, are provided online as supplementary material. If you are familiar with the R or Python programming environment, feel free to skip this chapter.

Introduction

Marketing Analytics and Data Science is ultimately a hands-on discipline: you learn by working directly with data, building models, testing assumptions, and communicating results. R and Python make this possible in a practical and reliable way. They allow you to import data from spreadsheets, databases, and

Supplementary Information The online version contains supplementary material, including data, Python code (R code is provided inline in the textbook) and outputs, which are available at https://doi.org/10.1007/978-3-032-11130-2_1.

© The Author(s), under exclusive license to Springer Nature Switzerland AG 2026
X. Dong, *Marketing Analytics and Data Science*,
https://doi.org/10.1007/978-3-032-11130-2_1

APIs; clean and reshape messy real-world datasets; visualize patterns and trends; and apply statistical and machine-learning methods at scale. In addition, programming turns data analysis into a repeatable workflow, so the results can be checked, updated with new data, and maybe reused for other similar projects without starting from scratch. In this textbook, we use R or Python not because marketing analytics or data science is "about coding," but because these tools provide the most flexible way to turn questions into evidence and evidence into decisions. In addition, both R and Python have large, supportive user communities, where valuable resources are shared freely online. These resources, together with the recent development in AI programing assistants, have dramatically reduced the learning cost for these languages.

Getting Ready for Programming Using R

R is an open-source programming language that is freely available for everyone to use. It is very popular in many fields for its capabilities in handling data analysis and creating data visualizations. Its popularity also encourages many users to contribute packages, facilitating more specialized uses and enhancing its reach to even more applications. These packages are all free to download from the Comprehensive R Archive Network (CRAN).

While the programming language is powerful, its user interface created a barrier for first-time users until the introduction of RStudio. RStudio provides a user-friendly integrated development environment (IDE) for R programming. It is also free and open-source.

Download and Install R and R Studio

To get started, you first need to install R and RStudio. They are available on the Posit website at https://posit.co/download/rstudio-desktop/. They support installations on multiple platforms.

Getting Started with R Studio and R Markdown File

RStudio provides a user-friendly interface for programming in R (Fig. 1.1). When we first open RStudio, it starts the software with three windows:

Fig. 1.1 RStudio logo

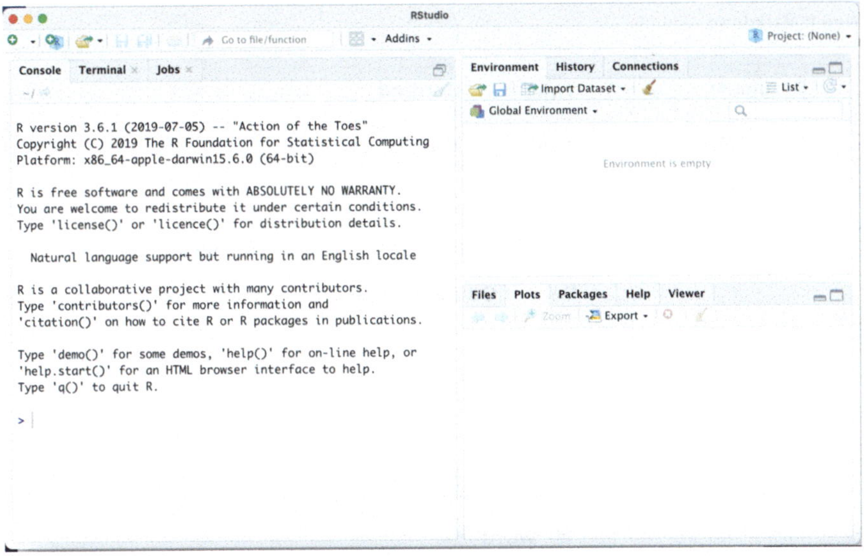

The one on the left is also called the Console, where we can begin typing our R commands. The two windows to the right are related to the outputs and the file locations.

RStudio allows us to write two types of files:

1. Traditional script file—R script: in which you can write the code, and it can be executed, and the results are going to be shown in the console below.
2. Notebook or Markdown file: these files combine the R script, execution results, texts, pictures, and equations into one file. In our example and homework questions, most of the files we will be using are Markdown files. After editing the file, we can knit it to a selection of file formats, including PDF, HTML, or a Word document. In our context, a Word document is the preferred format. To create a new Markdown file, we need to click on File → New File → R Markdown …

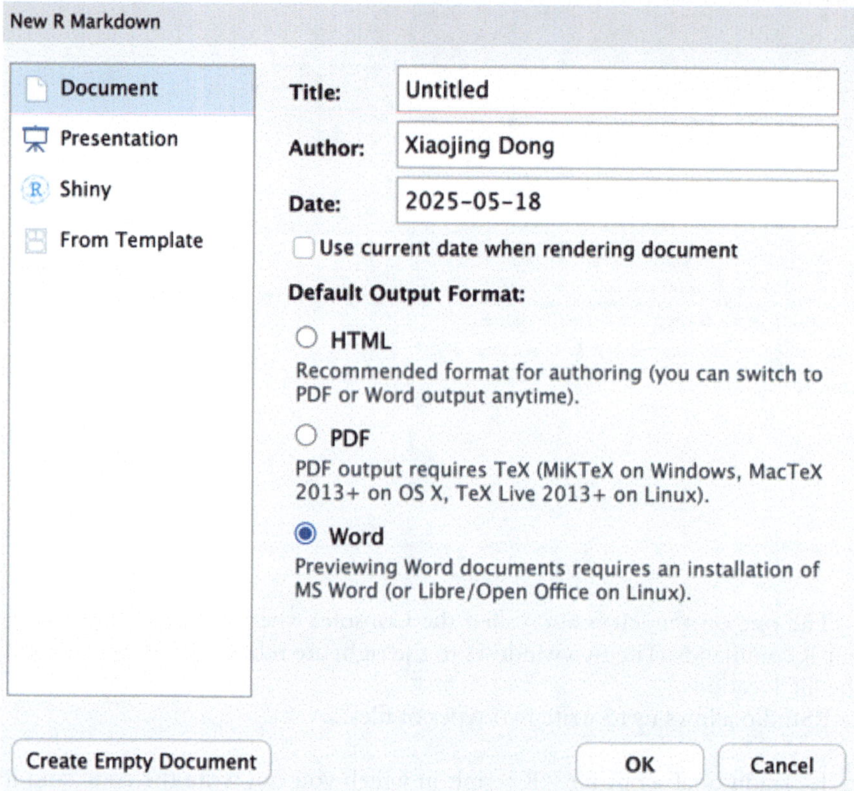

A window will pop up: In this window, you can specify the Title, Author Name, and Date. This information can be edited later, if needed. When you click OK on that window, a new Markdown file will be created for you to begin working on.

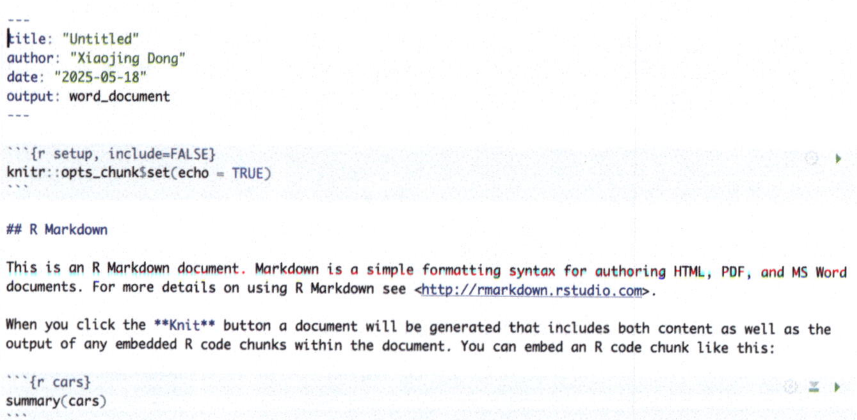

```
---
title: "Untitled"
author: "Xiaojing Dong"
date: "2025-05-18"
output: word_document
---

```{r setup, include=FALSE}
knitr::opts_chunk$set(echo = TRUE)
```

R Markdown

This is an R Markdown document. Markdown is a simple formatting syntax for authoring HTML, PDF, and MS Word
documents. For more details on using R Markdown see <http://rmarkdown.rstudio.com>.

When you click the **Knit** button a document will be generated that includes both content as well as the
output of any embedded R code chunks within the document. You can embed an R code chunk like this:

```{r cars}
summary(cars)
```
```

R Markdown File

Within this R Markdown file, the top area, bracketed with "---", is also known as the YAML (Yet Another Markup Language) format, which specifies information about the file in an easy-to-read language. You can easily edit and update the information, such as the title, date, and author.

The block with the shaded area, which is enclosed by "```", contains the R scripts that you are going to add. When RStudio encounters this format, it recognizes it as an R script and knows it needs to be executed using the R engine.

In the same shaded area, to the top right corner, there is a green triangle button. When you press the button, the R scripts in this shaded area will be executed. When the scripts are split into multiple blocks, a particular block cannot run unless the previous blocks have finished running. To do that, we can press the button to the left of the triangle button, that will execute all the scripts from the beginning of the file up to the block immediately preceding the current one.

Outside of the shaded blocks are textual contents. It is similar to editing a Word document, except you will need to specify the fonts or format.

For example, the text right after "#" indicates the top-level title, right after "##" indicates the second-level title, and so on. You can also bold the text by enclosing it with "**".

While the R Markdown file format includes more specifications, most of the homework you are dealing with will already have them laid out.

Once the file is finished editing, you can convert it to a distributable format, such as PDF, HTML, or a Word document. To do that, you can click on the Knit button and select the file format you want to knit into from the drop-down list. For this textbook and homework, you can knit everything into a Word document.

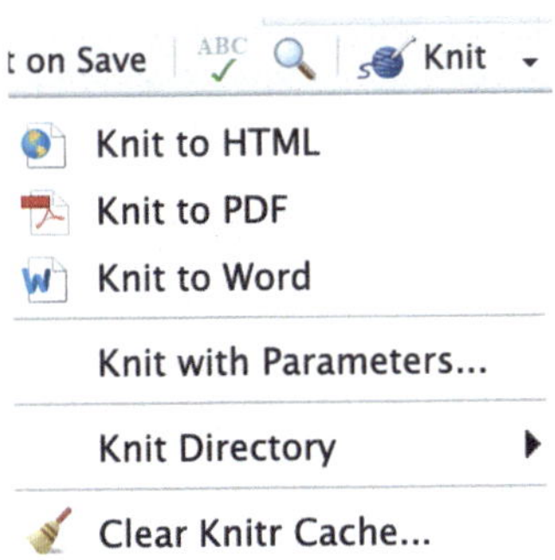

An Example of an R Markdown File

Using the CoffeeData.csv dataset, we illustrate how to get started with R programming.

First, Set Working Directory

It is convenient to keep all relevant files and data in a single folder. It allows the R code to access any files stored there, and any outputs the code generates can be saved into the same place. This folder is called the "working directory". So, the first step is to tell RStudio which folder is the working directory.

There are two ways to specify the working directory for R.

First method: Use the RStudio drop-down menu

To do this, at the RStudio window, we need to go to the small window on the bottom right, which shows multiple tabs. The first tab shows "Files",

1. Click on the "Files" Tab, as highlighted with number 1 in the screenshot, the folder in the current directory will be shown below.
2. Click on the "...", as highlighted with number 2 in the screenshot, and a "Go To Folder" window will pop up that allows you to select a folder.

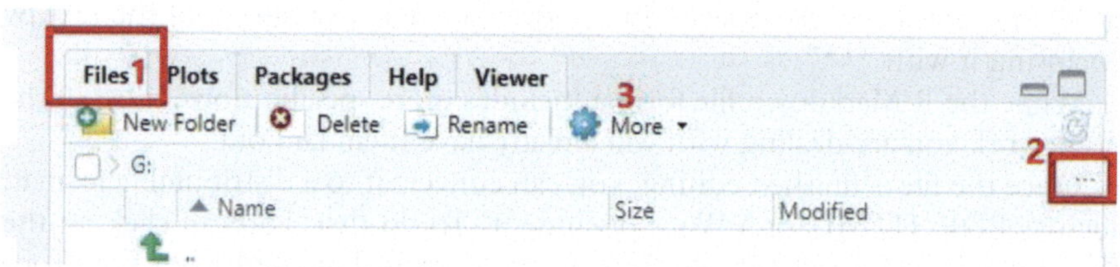

After selecting the folder where you stored the data files and the relevant files, we can set it as the working directory.

3. Navigate to the folder you plan to use as the "working directory", to set the current folder you are at as the working directory, click on "More" to see a drop-down list (see the figure below), and select "Set As Working Directory".

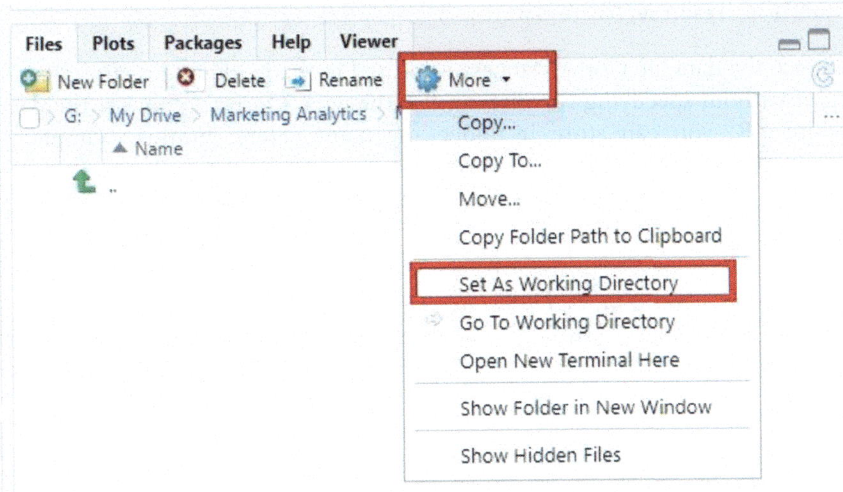

After that, you will notice that on the bottom left window, it shows the R command used

```
setwd("G:/My Drive/Marketing Analytics/Module 1")
```

Second method: Use the R command

The R command shown in the console window is the command that allows you to set the working directory. Therefore, you could use this command directly to set the working directory.

You can also check which working directory you are in right now using

```
getwd()
```

Here is the official R manual page about this R command http://stat.ethz.ch/R-manual/R-patched/library/base/html/getwd.html

Second, Read in the Data File

After setting the working directory, we can read the data file into R, allowing us to analyze the data using R. R can handle multiple data formats with similar commands. Throughout this textbook, the data files are in `.csv` format.

```
datafile <- read.csv('CoffeeData.csv')
```

Since the data file CoffeeData.csv is located in the working directory, R can find it without specifying the directory of the data. Once you import the data file into R, you can store it under any name. Here, in this example, I name it *datafile*.

Once the data is read into R, you can write R scripts to access and analyze the data.

Access Data Columns in the Data File

The data file consists of multiple variables, with one variable in each column. To start working with the data file, it is helpful to view the variable names first by using

```
names(datafile)

## [1] "day"              "dayofweek"        "Sales"            "Price"
## [5] "PromotionIndex"
```

To calculate the correlations between the variables *Sales* and *Price*, we can use

```
cor(datafile$Sales, datafile$Price)

## [1] -0.2060556
```

Create Some Simple Plots of the Data

To plot the *Sales* variable, we can use the command *plot()*, among which *type="l"* makes line plots.

```
plot(datafile$Sales, type="l", col="red")
```

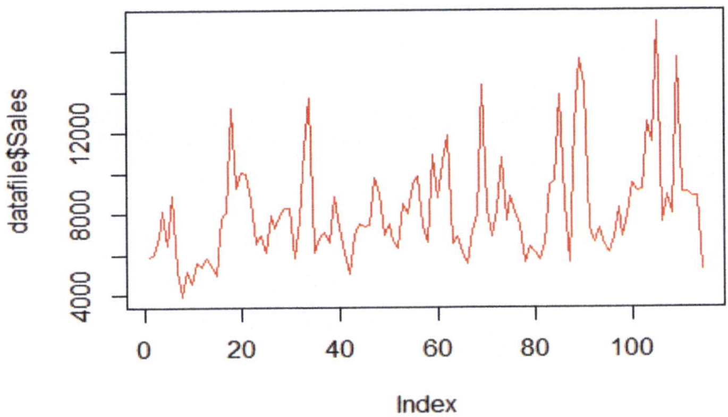

To plot the ***Sales*** variable together with the ***Price*** variable in the same figure, we can still use the command ***matplot()***

```
matplot(cbind(datafile$Sales,datafile$Price),type="l",
       lty = 1:2, col=c("red","blue") )
# lty specifies the different types of lines, 1: solid, 2: dotted

# adding the plot legend to specify the different lines
legend("topleft", c("Sales","Price"),
```

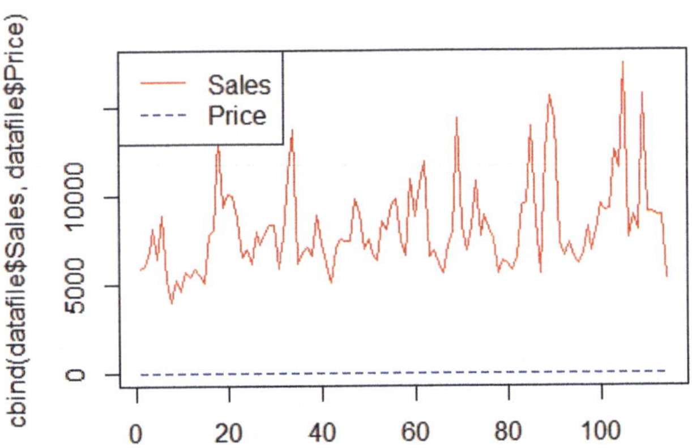


```
       lty = 1:2, col=c("red","blue") )
```

One issue with the above plot is that it is almost impossible to see the price variations. A closer examination of the values reveals a significant difference in scale between the two variables. To accommodate this, we can rescale the Sales variable before plotting them together.

```
matplot(cbind(datafile$Sales/1000,datafile$Price),type="l",
        lty = 1:2, col=c("red","blue") )
# lty specifies the different types of lines, 1: solid, 2: dotted

# adding the plot legend to specify the different lines
legend("topleft", c("Sales/1000","Price"),
        lty = 1:2, col=c("red","blue") )
```

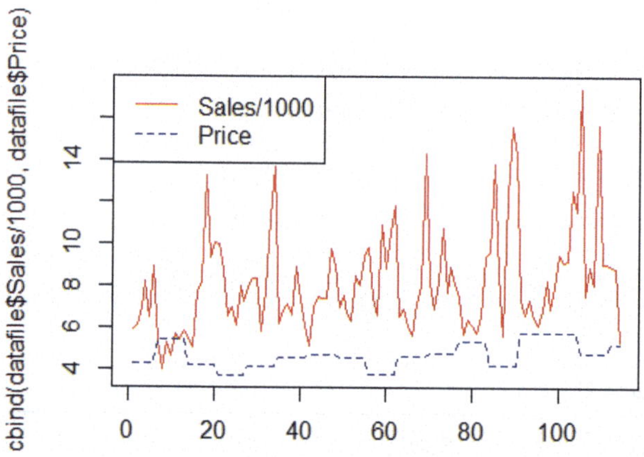

Get Packages

One of the reasons that R has become so popular in recent years is the extensive collection of R packages created by the R user community. A package typically contains functions, data, and compiled code, along with a user manual that provides detailed information about the functions included in the package.

You can find a complete list of R packages at https://cran.r-project.org/. It is not recommended that you go through the complete list of packages unless necessary.

To obtain a package, you will first need to learn its name, then use the function `install.packages("name_of_package")` to install the package and the relevant files onto your computer. Throughout this textbook, we will introduce a few additional packages, in addition to the default ones that are already installed when R is installed.

Get Help

There are generally two ways to get help on using R. If you know the name of the function that you are using, but would like to learn about its details or parameter definitions, use **help()**, or?.

```
?plot
help("plot")
```

Either way, you will get the official R manual for the function `plot()`.

If you are unsure of the function to use, the best way to find help is through an online search. R has become so popular that online communities are providing really useful suggestions and instructions to R users, for free. Below are a few websites that you might find helpful in getting answers to many of your R programming-related questions.

- https://stackoverflow.com/ is a community for coders. Not just R, for any programming language, if you have questions, you will find them answered by the community soon.
- http://www.statmethods.net/index.html is a well-organized website that helps beginners get started with R.
- https://blog.rstudio.com/ is the RStudio official blog site.
- https://www.r-bloggers.com/ is the R blog site.

In addition, here are a couple of books that may be useful:

- If you want to read a book about R, here is the link to the PDF version of The Art of R Programming.
- Here is another book that is useful to learn the basics for R graphics: R Graphics Cookbook: Practical Recipes for Visualizing Data.

Getting Ready for Programming Using Python

The Python programming language has become increasingly popular across various fields. Similar to R, its popularity is also fueled by the free resources provided by the growing Python community.

Before programming with Python, one needs to

- Install Python
- Install an Integrated Development Environment (IDE) for Python programming

Download and Install Python

There are multiple ways to download and install Python.

Download from the Official Website
The official Python can be obtained from its website for free
https://www.python.org/downloads/

Depending on the operating system of your computer, such as Windows or Mac, you can choose the correct version of Python to install.

Download and Install Anaconda
Alternatively, you can install Python using Anaconda. Anaconda is a widespread distribution of Python, which comes with many commonly used libraries (such as Pandas and Numpy), making it convenient for data analysis. The data science community widely adopts it as it provides an all-in-one environment for working with data.

It can be downloaded from https://www.anaconda.com/download (Fig. 1.2).

Once Anaconda is installed on your computer, you have all the Python resources needed to write and run Python code.

In the meantime, Anaconda Navigator also provides a selection of IDEs. One of the most popular IDEs is Visual Studio Code (VS Code), provided by Microsoft. You can click the download button from the Navigator, or it can be downloaded from this link https://code.visualstudio.com/download (Fig. 1.3).

VS Code is widely used for Python programming and can be installed without Anaconda. It is lightweight, and if you install VS Code without Anaconda, you can simply add the Python and data analysis packages later as you need them When you launch VS Code for the first time, you will see a window (Fig. 1.4):

Fig. 1.2 Anaconda logo

Fig. 1.3 Visual Studio Code logo

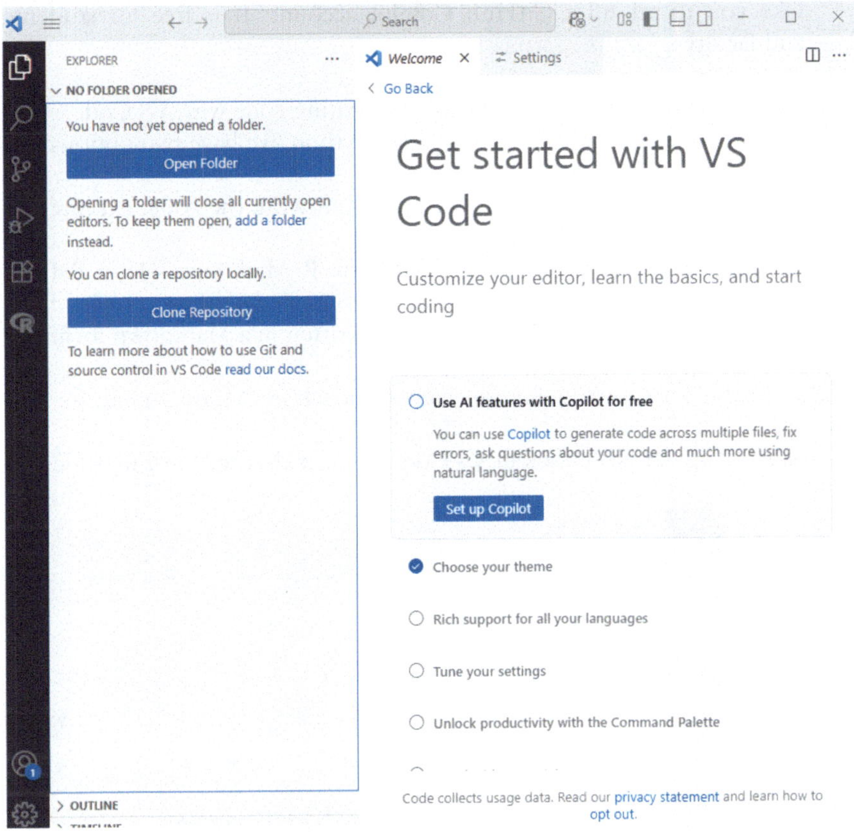

Fig. 1.4 Visual Studio Code launch window

Among the blue buttons:

- Open Folder: allows you to go to a folder on your computer where you want to save your Python code, and where you have your data stored.
- Clone Repository: allows you to connect with your GitHub account and sync a repository, making it easier to collaborate on a Python project with others. Since this step is not necessary for starting Python programming, we will not elaborate on the details here. For interested readers, please feel free to explore the benefits of GitHub Repositories for project collaboration and version control. You can visit https://github.com/ to learn more.
- Set up Copilot: provides an AI-assisted programming experience. It is highly recommended for first-time learners. Clicking on the button will

take you to set up a GitHub Copilot account. It is free for students and faculty.

Once the software is set up, you can start writing code with VS Code. Similar to the case in R, when we code in Python, we have two options:

- Traditional Python script file, where we can write the code, execute the code, and obtain the results.
- Jupyter notebook, which is similar to the R Markdown file, a Python notebook file allows the code to be combined with text. The texts included in a Jupyter notebook are also written in a Markdown format.

The Jupyter notebook is the preferred approach in Python coding for this textbook.

To start a Jupyter Notebook in VS Code, we can choose "Start new file…".

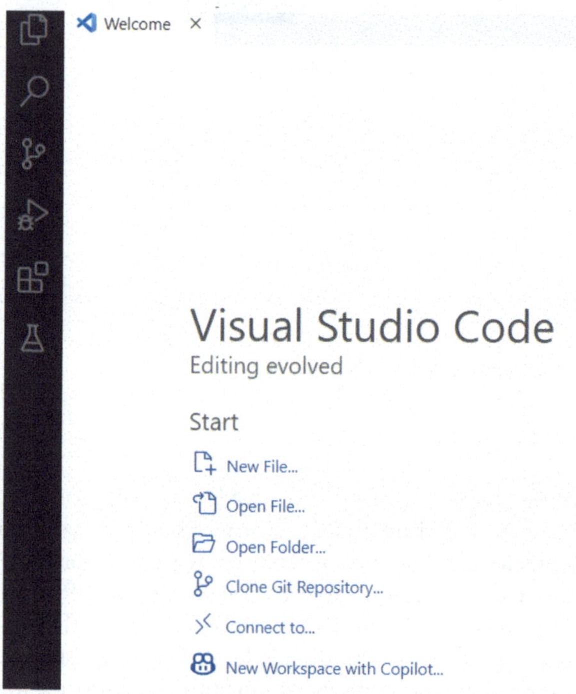

When we first start a Jupiter notebook, it also shows a shaded block.

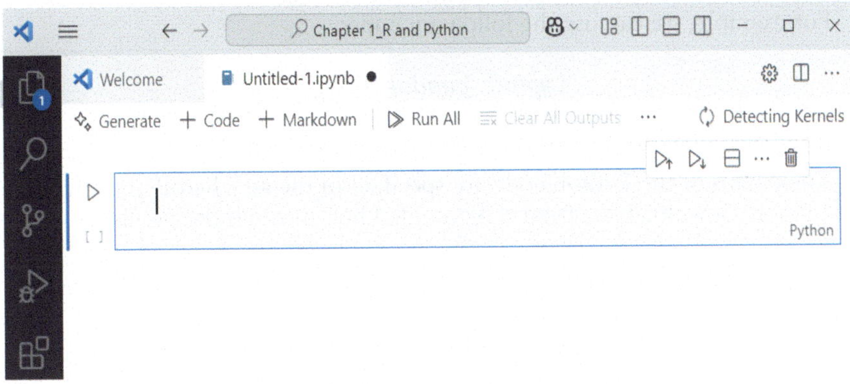

At the bottom right corner of the shaded area, it shows "Python", indicating that this block entails Python code.

To add text, you can add another block for Markdown, as indicated by markdown at the bottom right of the block. The syntax of editing texts in the markdown block is the same as that in editing an R Markdown file.

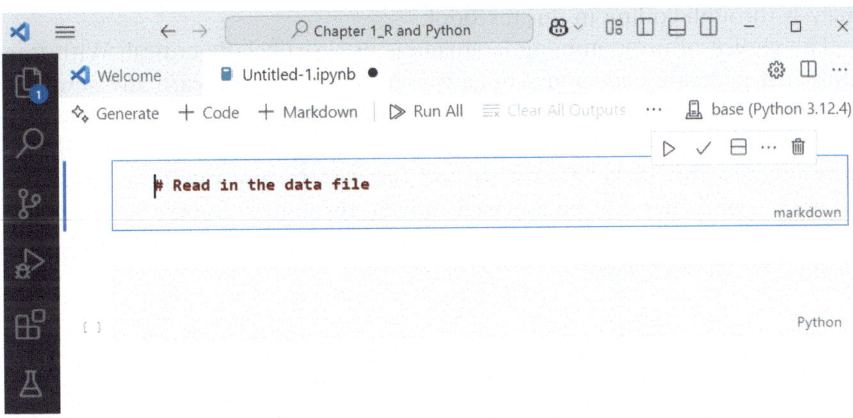

An Example to Use Python Jupyter Notebook

Very similar to R, Python also offers a wide range of packages available for download. One of the most popular packages is *Pandas*, which is a Python data analysis library. Unlike in R, after importing the library, Python must specify the library's name in front of every function provided by the library, separated by a dot, each time it uses a function from the library.

For example, when we need to read a CSV file into a Python DataFrame for data processing, we use the read_csv() function from the *Pandas* package. To allow the code to understand the function, the name of the pandas library is also required. This may be a hassle, especially when such a command is used multiple times. To reduce the effort of typing and to make the code easier to read and comprehend, we often take a shorter name.

For example, we can use the following code:

```
import pandas as pd
```

After this claim, we can use `pd.read_csv()` in the code.

The details of the Python code are specified in the attached Python Jupyter Notebook, downloadable from the provided link through the publisher.

Summary

This module helps you to get started with the Marketing Analytics course, through

- Setting up R and R Studio and understanding its basic commands
- Setting up Python with Anaconda or VS Code and learning its basic concepts and useful commands

These steps are essential before we can begin our journey of learning data analysis through coding in this textbook.

The choice of programming language is not particularly critical. With free tools that generate code and scripts, it is relatively easy to learn any new language. Both R and Python are beginner-friendly and supported by large, active online communities that continuously create new packages and tools. These communities also make it easy to find answers to any questions you may encounter, and they can be accessed quickly through a Google search or by leveraging large language models, such as OpenAI's ChatGPT, Google's Gemini, or Meta's Llama.

Homework Questions

1. Please follow the steps in this chapter to set up your working environment, either in R or in Python.
2. Download the data file CoffeeData.csv, and execute the code in the book, either in R or in Python.
3. Check if you can get the same results as those in the book.

CHAPTER 2

The Marketing Questions and Data Science Tools

Learning Objectives

1. Define and review core principles of marketing.
2. Comprehend the concept of marketing analytics, and its connections between marketing theory and data science tools.
3. Distinguish between strategies for implementing customer acquisition and customer retention.
4. Review key concepts in basic statistics concepts relevant to marketing analytics.

Introduction to Marketing and Analytics

This chapter introduces the foundational ideas that connect marketing and data science and prepares readers for data-driven analysis in marketing contexts. It begins by defining marketing and emphasizing its focus on understanding customers and creating value. The chapter then explains data science as the process of drawing insights and inference from data, and clarifies how the two disciplines work together: marketing frames the business questions, while data science provides evidence-based answers through data collection, modeling, and analysis. To support this connection, the chapter reviews essential statistical concepts, such as mean, variance, and the normal distribution, that are foundations for many data-driven analysis used throughout the book.

Supplementary Information The online version contains supplementary material available at https://doi.org/10.1007/978-3-032-11130-2_2.

© The Author(s), under exclusive license to Springer Nature Switzerland AG 2026
X. Dong, *Marketing Analytics and Data Science*,
https://doi.org/10.1007/978-3-032-11130-2_2

What Is Marketing?

Marketing involves a comprehensive strategy employed by organizations to connect with their target customers. The goal of marketing is not just to drive sales but also to build brand awareness, establish and maintain a positive reputation, and foster long-term relationships with customers. More concisely, marketing is all about customers: understanding customers to provide products and services that satisfy customers' needs and wants.

At the core of any marketing decision is the understanding of the **3Cs**, including:

1. The Company: the goal of the company, its competitive advantages, its resources, and challenges.
2. The Competitions: the current and the future potential competitive environment.
3. The Customers: the selected customers to target, including both the potential customers and the existing customers. These may be the customers that both the company and the competition are fighting to win over (Fig. 2.1).

Traditional marketing textbooks describe the marketing mix as consisting of the famous **4Ps**, including:

- Price: the price of the products and services serving the targeted customers
- Promotion: how, where, and when to promote which product to the targeted customers
- Product: what products to offer, what product features should be included
- Place/channel: the best way of reaching the customers with the product and service, such as through resellers or through retailers (Fig. 2.2)

At the core of every marketing decision lies the customer. It is common to categorize marketing strategies into two types: customer acquisition and customer retention.

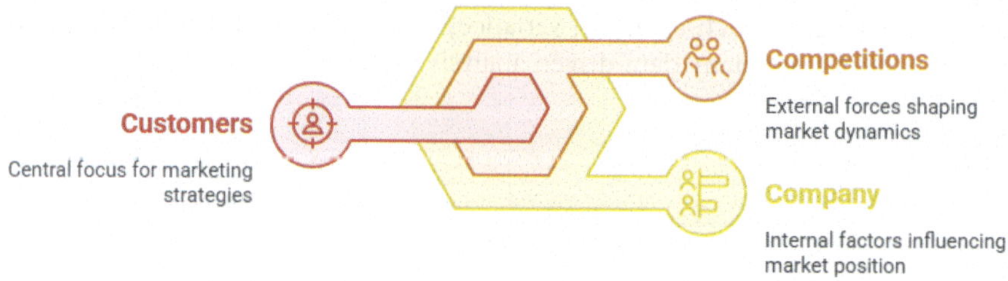

Fig. 2.1 The 3C framework. (Note: This picture is generated with napkin.ai)

Fig. 2.2 The marketing mix: 4Ps

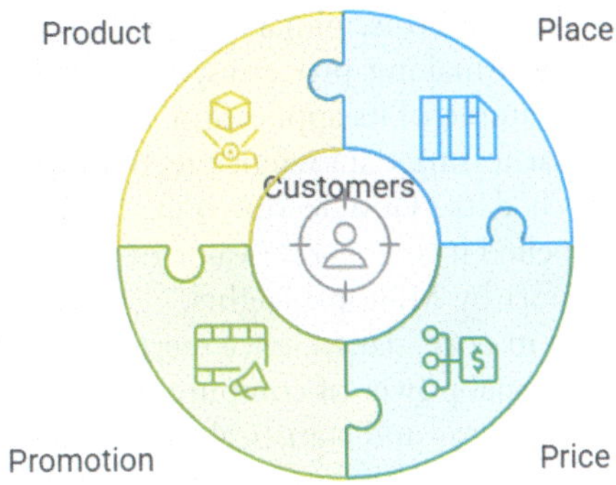

- Customer acquisition
  These strategies focus on identifying and converting potential customers into actual buyers, for both individual consumers and business customers. The objective is to grow the business by expanding its customer base and increasing market reach.
- Customer retention
  These strategies are centered on maintaining relationships with existing customers. The goal is to foster loyalty, encourage repeat purchases, and maximize the lifetime value of each customer.

In a customer-centric marketing approach, it is essential to understand customer behavior, their decision-making processes, and the key factors that influence their decision-making processes. Developing this understanding depends on the collection and analysis of customer data. Traditionally, customer insights were gathered through direct interactions with sales personnel. While this method can provide rich, personalized information, it lacks scalability and is susceptible to knowledge loss when employees leave the organization.

To address these limitations, businesses increasingly rely on approaches that can systematically capture and manage customer information in digital form. By storing customer data in organized databases, companies ensure that insights are preserved, accessible across teams, and scalable as the customer base grows. This data-driven approach enables marketers to make informed and efficient decisions.

What Is Data Science?

Data science has emerged as a prominent field around the early 2010s, offering a suite of analytical tools that have transformed how we interpret data. Originating at the intersection of computer science and statistics, data science now extends its influence across diverse fields, from social sciences and natural sciences to practical applications in the arts, politics, human resources, healthcare, automotive technology, finance, product development, and beyond.

This interdisciplinary field has been instrumental in refining business decision-making processes, particularly in marketing and customer analytics. An example of its application is Google's pioneering role in search advertising. By harnessing data science techniques to mine insights from vast quantities of search data, Google not only created the search advertising sector but also propelled it to a market value of around \$164.12 billion by 2020, according to a report by Million Insights.

At its core, data science merges the analytical rigor of statistics with the computational power of computer science. Statistics provides the methodology for data analysis and statistical inference, allowing for the construction of models that reflect the complexity of the data generation processes. The practical use of these models depends on computational tools, which have become increasingly accessible through advancements in programming languages such as R and Python. These languages, supported by a rich ecosystem of code libraries and the recent development in large language models (LLM), have dramatically lowered the effort required for learning programming and data analysis.

In this book, we aim to equip the readers with the fundamentals of R and Python programming, essential for delving into data science. Alongside detailed code examples and explanations, we offer links to further online resources. This comprehensive approach is designed not only to equip the readers with essential tools but also to encourage exploration and foster deeper learning within the dynamic data science community.

What Is Data Science and Marketing?

At the intersection of data science and marketing, marketing poses the questions, and data science supplies the solutions. These marketing questions often revolve around understanding customers and their decision-making processes and behaviors. This pattern is evident in both individual consumers (business-to-consumers, or B2C) and business organizations (business-to-business, or B2B). Given the vast and rapidly expanding datasets associated with customer information, data science and artificial intelligence tools become essential for developing models through sophisticated computational techniques.

To navigate data science tasks for informed marketing strategies, proficiency in several key areas is required:

- Marketing: Insight into marketing principles is crucial for identifying relevant questions and translating business objectives into data-driven inquiries.
- Economics: This discipline provides a foundation for understanding the rationale behind consumer and organizational decision-making, which is important for constructing relevant models.
- Statistics: Essential for all forms of data analysis, statistics enable the drawing of meaningful inferences from complex datasets.
- Computer Science: Offers the computational techniques necessary for handling and analyzing large volumes of data (Fig. 2.3).

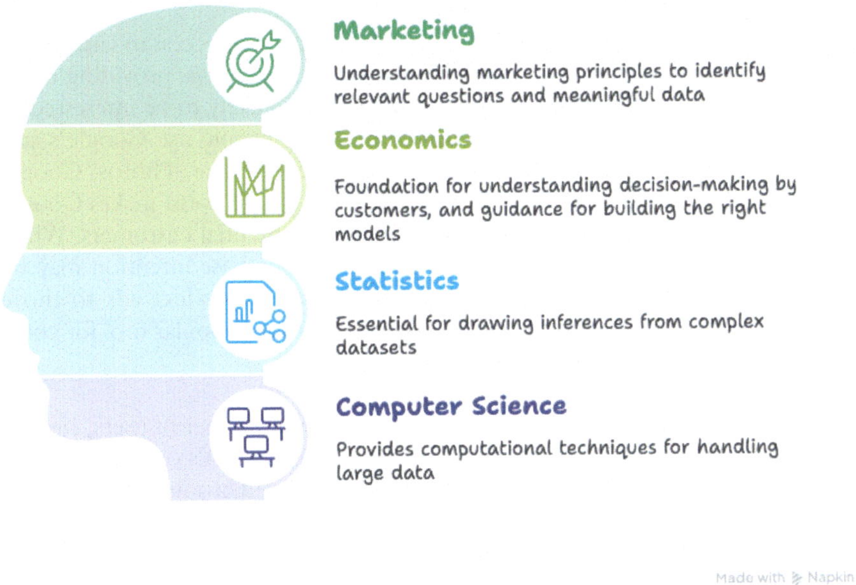

Fig. 2.3 Key areas required for implementing data science in marketing

Two Types of Tasks for Analyzing Customers

To implement data-driven marketing strategies and analyze customers' decisions, we can categorize them into two types of tasks: customer acquisition and customer retention.

Customer Acquisition
Acquiring new customers presents two significant challenges in leveraging data science tools:

- First, identifying promising prospects from a large population faces the challenge of having access to only limited initial data.
- Second, converting those prospects into customers requires reaching them first; however, tools to reach non-customers may not be readily available.

Traditionally, this process relied on resource-intensive methods such as cold calling, sales visits, and attending trade shows. However, the digital era has revolutionized this process by introducing more scalable and cost-effective tools.

Online platforms, such as Google, Facebook, and LinkedIn, now enable businesses to reach broader audiences at a very low cost. The customer acquisition process typically involves two stages:

- First, Identifying Access Channels
  Businesses can leverage digital platforms that provide access to large pools of potential customers. These platforms act as gateways, providing companies with visibility among users who are more likely to be interested in their products or services. For example, many people use Google's services daily at no cost, such as Gmail, Google Calendar, Photos, Google Drive, and more. The large free audience on the platform makes Google a default place when a business tries to reach potential customers. When people search for products on Google, their purchase intention may be revealed. Therefore, Google can display relevant product ads to those users. As a result, Google search ads became a very popular tool for companies aiming to acquire more customers.
- Second, Engaging and Converting Prospects
  Using data available on the platforms with large number of users, such as Google or Facebook, businesses can identify individuals or organizations who closely match their ideal customer targets. Through targeted and personalized advertising, companies can increase the effectiveness of their outreach efforts and boost conversion rates.

The enhanced efficiency of digital customer acquisition stems from the strategic usage of diverse data sources and advanced targeting technologies. Together, these tools enable marketers to optimize both reach and relevance, reducing waste and improving return on customer acquisition efforts.

Customer Retention
Once a prospect becomes a customer, businesses can begin to build a more complete customer profile using the data generated through these ongoing interactions. This customer-level information can be organized into two categories:

- Static Information
  This includes demographic details for individual consumers (e.g., age, gender, location) or firmographic data for business clients (e.g., industry, company size, company revenue). These attributes are generally stable over time and provide a foundational understanding of the customer profiles.
- Dynamic Information
  This captures real-time customer behavior and engagement. For example, businesses with an online presence can track users' activities, such as browsing products, writing reviews, clicking ads, or abandoning shopping carts. These interactions provide valuable insights into customers' evolving preferences and needs. By analyzing both static and dynamic data, businesses gain a deeper understanding of customer journeys, anticipate future behaviors, and design retention strategies that foster long-term engagement and loyalty.

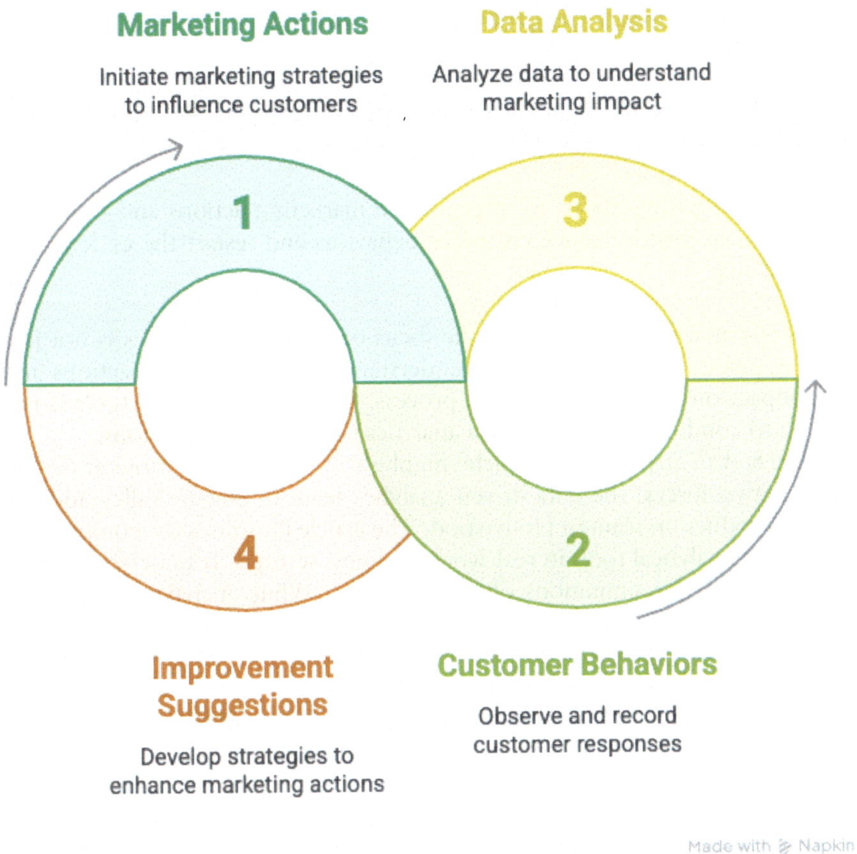

Fig. 2.4 The marketing analytics cycle

The Cycles of Marketing Analytics

Marketing analytics integrates data science tools to address marketing-specific inquiries. Expertise in marketing ensures the formulation of pertinent and meaningful questions, whereas data science techniques find solutions by extracting insights from within the data.

It involves two processes in an iterative process, as described in the following picture (Fig. 2.4):

The first process, from left to right, describes the data generation process, which involves:

- First, marketing actions from the company will influence customers' behaviors.
- Second, these customer behaviors are recorded in the data, together with the details of the customer interactions with these marketing actions.

The second process, from right to left, describes the marketing analytics process, which involves:

- Analyzing the relationship between the marketing actions and customer behaviors to uncover the influence of the marketing actions on customer behaviors
- Providing suggestions to improve the marketing actions and decisions, which in turn influence customer behaviors and restart the cycle to the first step

It is essential to recognize that the focus of marketing analytics is not just applying analytics tools, but rather understanding the marketing actions and their impact on the data generation process. Simply knowing the tools is not enough to conduct meaningful data analytics for marketing decisions.

A 2018 *Wall Street Journal* article[1] highlights the internal tensions at Netflix between two forces: the data-driven analytics team in Silicon Valley and the creative production team in Hollywood. The article illustrates the complexities of applying analytical tools in real-world business settings. It underscores both the power and the limitations of data analytics. While analytics can provide valuable insights, they are inherently constrained by the scope and quality of the underlying data. Many important business decisions involve factors that may fall outside of what was captured in the data. Therefore, it is essential not only to analyze the data but also to critically understand how the data was generated and what it may be missing for any particular decisions. Only then can analytical results be appropriately interpreted and effectively applied to real marketing and business decisions.

Review of Basic Statistics

Let's review some of the basic statistics we learned before, using R and the same dataset from the last chapter.

```
datafile <- read.csv('CoffeeData.csv')
```

Using `the ColMeans` function, we can calculate the mean value for each column in the dataframe.

[1] Ramachandran, S. and Fint, J. (2018, Nov. 10). "*At Netflix, who wins when it's Hollywood vs. the algorithm?*" *The Wall Street Journal.*

```
colMeans(datafile)
##           day     dayofweek          Sales          Price PromotionIndex

##     57.500000      3.973684    8170.263158       4.661404      19.349374
```

We can also use the function *apply*, which applies each column or each row in the dataframe to the function specified in *apply*.

```
apply(datafile, 2, var) #calculate the variance for each column (as indicated
 by value 2) in datafile

##           day     dayofweek          Sales          Price PromotionIndex

##  1.092500e+03  4.008151e+00   6.178318e+06   3.859237e-01   6.191126e+02

apply(datafile, 2, median)

##           day     dayofweek          Sales          Price PromotionIndex

##     57.500000      4.000000    7620.500000       4.580000       1.039346
```

Q: How is the variance calculated?

A: You need to calculate the mean first. Variance measures the average spread or average distance from the mean.

$$\text{var}(a) = E\left[(a-\mu_a)^2\right]$$

- To obtain the expectation, we calculate the average.

$$\mu_a = \frac{\Sigma_{i=1}^{N} a_i}{N}$$

- To obtain the variance, we need to know whether it is the population variance or the sample variance, as the denominators used in the calculations are different.

1. Sample variance $\text{var}(a) = \dfrac{\sum_{i=1}^{N}(a_i-\mu_a)^2}{N-1}$

2. Population variance $\text{var}(a) = \dfrac{\sum_{i=1}^{N}(a_i-\mu_a)^2}{N}$

N represents the sample size, which is the number of elements in the variable *a*. If $N \rightarrow \infty$, $N = N - 1$, then, the sample variance is equivalent to the population variance. In other words, if the size of the data is really large, the difference between the population and sample variances is negligible.

Q: How is the median calculated?

A: If the number of values included in the variable is an odd count, sort all the values; the one in the middle is the median value for the variable. If the number of values included in the variable is an even count, after sorting the values, select the two middle values, and then calculate the average of these two numbers. This average serves as the median for the variable.

Q: What's the difference between mean and median?

A: When calculating the mean, all the values in the data are utilized. However, when calculating the median, only the middle value or the two middle values are used after sorting the variable. As a result, if there are outliers, either extremely large or extremely small, their values will contribute to the calculation of mean not the median. Therefore, the median is unaffected by the outliers in the data. This is an important distinction and highlights the reason that in many applications using income values of a region, more often.

Covariance and Correlation

Calculate the covariance between two variables, a and b.

$$\text{cov}(a,b) = E\left[(a-\mu_a)\times(b-\mu_b)\right] = E(ab) - \mu_a\mu_b$$

It is calculated as the difference between the mean of the product of a and b and the product of the means. It may be easier to remember it as the mean of the product minus the product of the means.

Note: Usually, when we calculate covariance, it refers to "population" covariance.

Correlation can be considered as the scaled covariance, and it is calculated as

$$\text{cor}(a,b) = \frac{\text{cov}(a,b)}{\sqrt{\text{var}(a)\text{var}(b))}}$$

We will frequently encounter the concept of correlation throughout this course. Correlation refers to the statistical relationship or co-movement between two variables. When two variables exhibit a positive correlation, it means that as one increases, the other tends to increase as well. Similarly, a negative correlation implies that as one variable increases, the other one tends to decrease, as they move in opposite directions.

Correlation is a fundamental concept in data analysis because it helps identify patterns and relationships between variables. However, one of the most common pitfalls in interpreting correlation is to mistake it for causality, in which one variable is said to cause the other. For example, this confusion often arises in the context of regression analysis. A regression model requires the selection of a dependent variable Y and one or more independent variables X. While the model estimates the relationship between Y and each X, it does not imply that changes in *X* **cause** changes in *Y*, instead, it can only be interpreted as correlation between them.

This distinction between correlation and causation is critical and will be explored further in the chapter on linear regression. If you're interested in reading more about this topic, but in a more accessible way, feel free to explore the article Dong and Heneike (2016), which I co-authored with a colleague, Dr. John Heneike. It discusses the implications of mistaking correlation for causality in financial and business analytics.

Normal Distribution

Before reviewing the details of the normal distribution, let's quickly review the concept of a probability distribution.

In the statistics reviewed above, for example, the mean value of a variable *X* provides us with some basic statistical features of the variable. For example, when describing the average GPA (grade point average) of students from two schools, if we find that mean(X_1) > mean(X_2), it indicates that the students from the first school, with their GPAs represented by the variable X_1, on average is higher than that from the second school. This information is helpful, but it does not mean that all the students in the first school have higher GPAs than all of those in the second school. To understand how many students have higher GPAs, or the number of students at each GPA value, we need to know more information about the students' GPAs from each school. That information can be provided statistically through the distribution of GPAs for each school.

In general, the probability distribution of a variable provides far more information about the possible values of the variable than summary statistics such as mean and variance. A probability distribution describes how likely the variable is to fall within any range of values, while the mean and variance offer only a high-level summary, capturing the overall average and the spread of the possible values, but not the overall shape of the data.

The Normal distribution was first described by the French mathematician Abraham de Moivre in his book "*The Doctrine of Chances*", published in 1938. In this book, he showed that when a fair coin is tossed many times (independent trials, with equal probability of success), the distribution of the number of successes can be approximated by a bell-shaped curve, an early form of what is now called the Normal distribution. Several decades later, another famous French mathematician, Pierre-Simon Laplace, generalized this idea by showing that if we draw multiple large samples from a distribution and calculate the sample average for each sample, the collection of all these observed sample averages can be approximated by a Normal distribution. This is true even when the original distribution was not a Normal distribution. This is what we now know as the central limit theorem. It also establishes the theoretical foundation for the widespread adoption of the Normal distribution in data science.

The normal distribution can be illustrated in the two plots below.

```
#dev.new(width = 4, height=3, unit="in")
x = seq(-3,3,0.1)
def.type = par(mfrow=c(2,1))
y=dnorm(x)
plot(x,y,type="l",col="red",
     main = "Probability Density Function for Standard Normal Distribution",
     xlab = " ", ylab = "PDF values")
plot(x,pnorm(x),type="l",col="red",
     main = "Cumulative Distribution Function for Standard Normal Distributio
n",
     xlab=" ", ylab="CDF values")
```

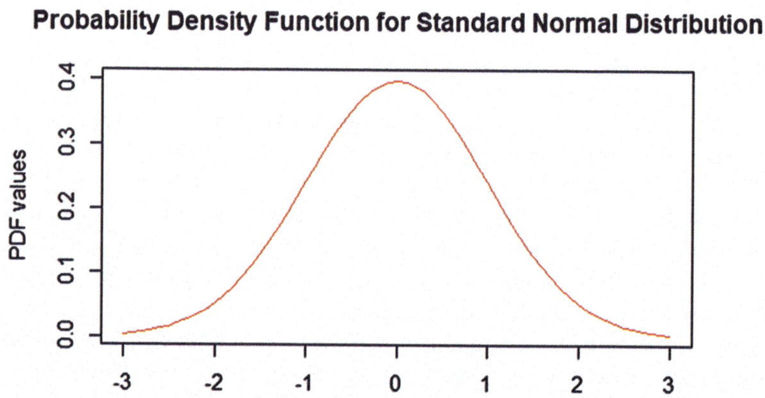

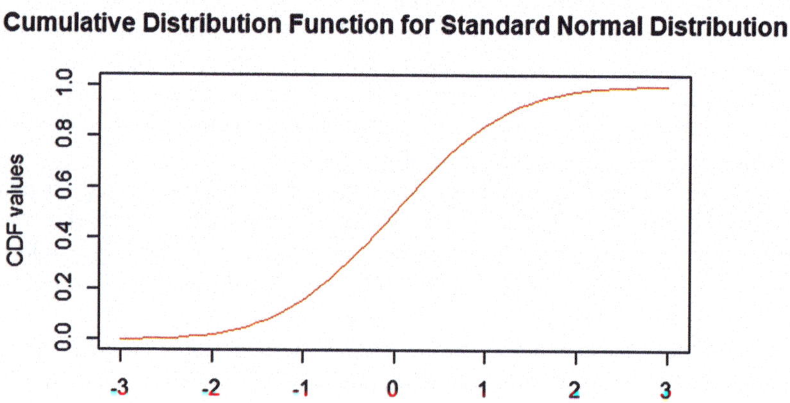

The top plot shows the probability density function (PDF) of a standard normal distribution. $X \sim N(0,1)$, and the bottom one plots the cumulative distribution function (CDF) of the same normal distribution.

Normal distribution has only two parameters:

- μ is the mean in the normal distribution, indicated by the center in the PDF plot.
- σ^2 refers to the variance in the normal distribution, indicated by the spread of the PDF plot.

The PDF for a normal distribution with parameters (μ, σ^2) can be specified as

$$f(x) = \frac{1}{2\pi\sigma} \exp\left[-\frac{1}{2}\frac{(x-\mu)^2}{\sigma^2}\right]$$

The CDF function calculates the area below the curve, up to the value X.

$$F(X) = \int_{x=-\infty}^{x=X} f(x)\,dx = \int_{x=-\infty}^{x=X} \frac{1}{2\pi\sigma} \exp\left[-\frac{1}{2}\frac{(x-\mu)^2}{\sigma^2}\right] dx$$

Normal distribution (PDF) has a few essential features:

1. It is also called the bell curve, due to its shape.
2. It is symmetric.
3. It has only one mode, that is, only one peak, as illustrated in the above plot.
4. It has relatively long tails on both sides.
5. Its Cumulative Distribution Function (CDF) does not have a closed-form solution.

These features also remind us that the Normal distribution is a very useful model, but only in settings where the assumptions are reasonably satisfied. Its assumptions can fail in some applications, and in rare cases, the failure can be consequential. Despite this, the Normal distribution is widely used because it is mathematically convenient and often provides a reasonable approximation for many everyday phenomena. Therefore, it is important to understand what can go wrong when a Normal assumption is applied to contexts where extreme outcomes are more common than the model suggests. As a prominent example, Nassim Nicholas Taleb argues that many financial risk models relied too heavily on Normal assumptions and therefore underestimated rare but high-impact events. This underestimation can create a false sense of safety and leave systems fragile when extreme "black swan" shocks occur (Taleb, 2007). He cited the 2008 Financial Crisis is one prominent example.

Summary

This chapter helps you to get started with the marketing analytics course, through

- Introducing the fields of data science and marketing, and their intersection for marketing analytics, where data science tools and techniques are applied to marketing problems
- A brief review of the basic concepts of statistics, which we will encounter later in the course

Homework

Discuss an experience you have had with a company's product or service. Describe the type of data you believe the company has collected about you, and explain how they may use this data to encourage your repeated interactions with them and retain you as a long-term customer.

Reference

1. Xiaojing Dong and John Heineke (2016), "Correlation or Causation: The Sorry State of Inference in Empirical Modeling," Journal of Investment Management, 14 (3), pages 62–73..

CHAPTER 3

Unlocking Marketing Insights with Free Data

Learning Objectives

1. Explain the framework and key components of a customer data infrastructure.
2. Identify and collect curated external data from online resources when necessary.
3. Apply Application Programming Interfaces (APIs) to retrieve data from websites and arrange them in a format ready for modeling analysis
4. Recognize the importance of customer data protection and comply with relevant regulations when analyzing customer-level data.

Introduction

In the digital era, data has become the cornerstone of marketing decision-making. As highlighted in Chap. 2, data science and analytics tools have become essential for organizations to understand customer behaviors and to shape their marketing strategies. However, acquiring high-quality data can be expensive and complex. Fortunately, a growing number of free and publicly accessible data sources allow businesses and researchers to conduct meaningful marketing analytics without a significant investment.

This chapter introduces a framework for organizing marketing data, explores free tools such as Google Analytics for collecting internal data, and presents a rich set of external data sources available at no or limited cost. Additionally, detailed instructions are provided on how to gather and analyze these data using Application Programming Interfaces (APIs).

Supplementary Information The online version contains supplementary material available at https://doi.org/10.1007/978-3-032-11130-2_3.

© The Author(s), under exclusive license to Springer Nature Switzerland AG 2026
X. Dong, *Marketing Analytics and Data Science*,
https://doi.org/10.1007/978-3-032-11130-2_3

While customer data is critical to effective marketing, protecting customer privacy and ensuring data security are among the most important responsibilities for any organization that handles sensitive customer information. This chapter also presents an overview of global regulations and data protection policies. Organizations must adhere to these legal and ethical standards before collecting, storing, or analyzing customer-level data.

Building a Customer Data Infrastructure

Marketing analytics usually begins with internal data, collected through an organization's own digital properties and interactions with its customers. A robust infrastructure for organizing this data ensures long-term usability and scalability.

Internal data sources include web activity (such as page views, sessions, click-throughs, searches, and navigation), customer relationship management (CRM) systems, email interactions, social engagement, and transaction logs. Structuring these data into layers helps create clarity and potential future expandability:

- Raw Layer: Includes unfiltered logs and event data collected from various platforms embedded with tracking capabilities. These data tracks every action taken by a user or customer when interacting with the company,
- Cleaned Layer: Removes duplicates, corrects formatting errors, and standardizes data types,
- Analytical Layer: Aggregated summaries or visual dashboards used for reporting and monitoring user activities or business performances (Fig. 3.1)

To build effective customer analytics capabilities, organizations must track and store data in a systematic and structured manner. While some data are relatively easy to collect, others require more deliberate design and careful data tracking.

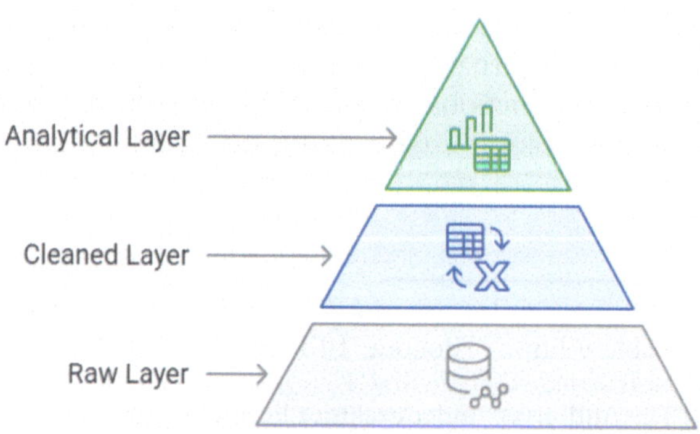

Fig. 3.1 Three-layer framework for company-owned data

For example, in the case of an online store, user behavior can be automatically tracked through event logs. These logs can capture a wide range of activities, including browsing history, page views, product clicks, purchases, and engagement with platform features such as customer service chat. They can also record whether a user clicked on a particular advertisement or purchased a product recommended by the system. These behavioral signals form the foundation for understanding user preferences, which in turn support personalization strategies that e-commerce can offer.

Despite careful tracking of user data, some important information may be missing.

In today's environment of dynamic targeting and algorithm-driven personalization, every user may be exposed to a unique set of advertisements and product recommendations. However, the information regarding the specific ads or products shown to each user is often overlooked or not systematically captured in the data. Failure to record this exposure information limits the ability to measure the actual impact of targeting and recommendation strategies from the available data. Capturing this layer of interaction is critical for evaluating marketing effectiveness at the individual level and unlocking the full potential of data-driven decision-making.

In general, tracking customer data is relatively straightforward, as it can be easily automated using first-party or third-party tools. More attention is needed to the records of company decisions and actions, especially those that directly influence customers' choices and decisions.

External Free Data Sources for Marketing Research

While internal data provides detailed insights into existing customers, external data can enrich the analysis by offering additional information on market trends, competitive intelligence, and insights into potential customers.

Some online sources provide access for a fee, while many others offer free, curated datasets that can be utilized to supplement internal data.

Below is a list of selected popular websites that offer freely available data sets:

- Google Trends
  Google provides line plots that illustrate how often specific search terms are entered in Google's search engine. These illustrations can be helpful for brand interest analysis and demand forecasting. It can also help to identify popularity trends. The data does not report the exact number of searches for each time period; it presents values in relative terms, scaled against the peak value, which is set to be 100% (Fig. 3.2).

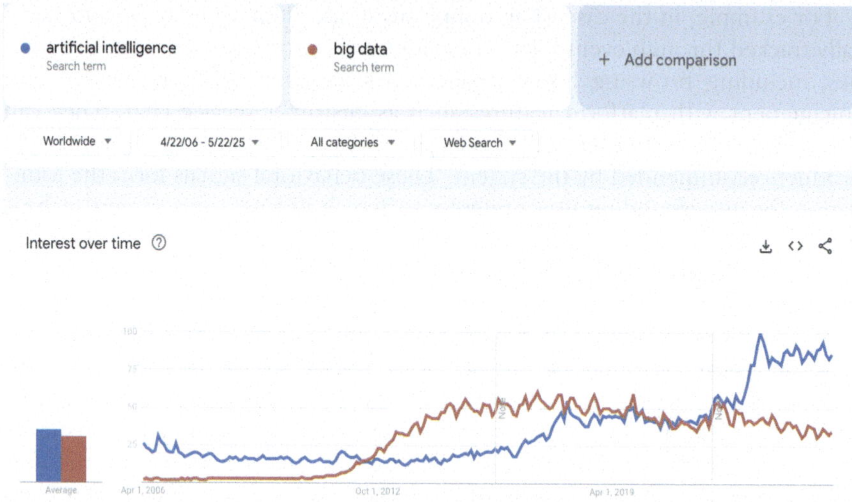

Fig. 3.2 Google Trends data illustration of keywords "artificial intelligence" versus "big data"

- Kaggle
  Kaggle.com was founded in 2010 as a community platform hosting data science competitions. In 2017, it was acquired by Google, which brought together the largest data science community with Google's machine learning cloud services. Today, Kaggle hosts a vast collection of datasets curated by individuals and organizations, all of which are freely accessible. The platform has significantly impacted the field of data science by promoting collaboration and learning.
- US Census Bureau
  The Census Bureau conducts a census survey every 10 years. This routine was initially established as a constitutional mandate to count the number of residents in each state, which determines the number of seats each state has in the US House of Representatives. The census data contains demographic information at the aggregate level. For example, they have information about the average income in a specific geographic location or a particular zip code. The data are organized and well-documented. This information can help evaluate the potential market size for a specific product or service.
- Yelp Open Dataset
  Yelp provides a dataset for free access, which contains business information, texts for user reviews, and pictures across multiple metropolitan areas. It can help analyze location-based sentiment or test labeling algorithms for texts and images using large language models.

These datasets can be downloaded directly from their websites. Additionally, numerous other datasets require API access or scraping, which are discussed in the next section.

Using APIs to Collect Data: An Example

When data is not readily available for direct download, some websites and data sources offer APIs that enable users to access data using computer programs. APIs (Application Programming Interfaces) enable the retrieval of structured data with proper authentication and data access limits imposed by the platforms.

For instance, if we aim to collect tweets and retweets related to a specific brand or its campaign on X.com (formerly Twitter.com), the platform provides APIs designed for such purposes. The general steps required to access this data are outlined below. Sample code in both R and Python is available on the platform's developer website, and readers are encouraged to consult that site directly. Given the frequent updates to API design and access policies, as well as corresponding changes in required R and Python packages, including sample code here in this textbook may risk becoming outdated before publication. Nonetheless, the overall workflow remains consistent, and the key steps and their purposes are explained below.

Step 1: Set Up a Developer Account on X.com

- Sign up for a developer account
  Begin by visiting https://developer.x.com/en to sign up for a developer account. A free-tier option is available, which is well-suited for learning and experimentation. While this tier imposes data access restrictions (which may change over time), it provides sufficient functionality for most educational and exploratory use cases.
- Create a new project and a new app within the project
  After signing in, you'll need to create a new project and an associated app. In some cases, the platform automatically generates a default project and app for you, which can be accessed via the dashboard on the developer portal.

If an app is not created automatically, you can manually create one by clicking the appropriate button in the dashboard.

- Generate API keys and tokens
  Once the app is created, as shown in Fig. 3.3, under "PROJECT APP" click the "Key" icon, which will take you to the "Keys and tokens" tab (see Fig. 3.4), where you can generate the keys and authentication tokens, which are necessary for API access.

Here is a brief overview of the key components (Fig. 3.4):

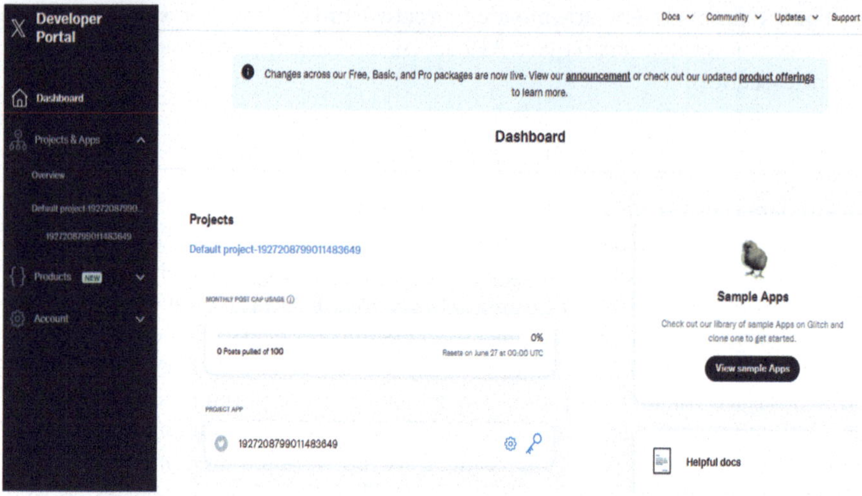

Fig. 3.3 Twitter dashboard on the developer portal

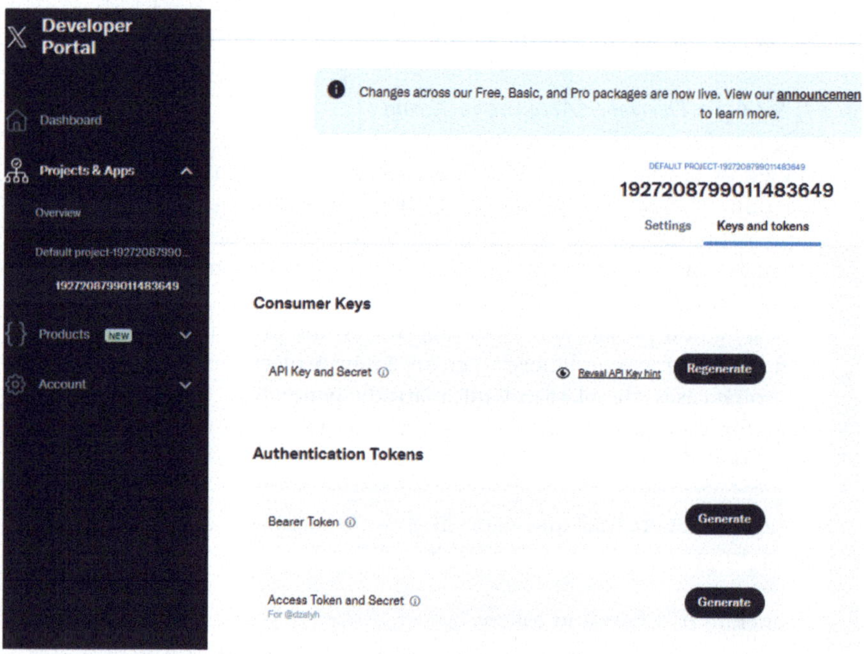

Fig. 3.4 Get authentication tokens from X.com

- **Consumer Keys**: These function like a username and password for your app and are required to authenticate API requests on behalf of your account. These keys should be kept secure and never shared or exposed in public code repositories.
- **Authentication Tokens**: These tokens validate and authorize your API access. Different tokens grant different levels of access:
  - **Bearer Token**: Provides read-only access to public data, such as searching for recent tweets. This is typically sufficient for most marketing analytics purposes.
  - **Access Token and Secret**: Used when user-level permissions are required (e.g., posting tweets on behalf of a user). These are generally not needed for read-only marketing analytics tasks.

Clicking the "Generate" button next to "Bearer Token" will open a pop-up window displaying a lengthy string, which is the bearer token required for the coding in the next step. Please copy the long string and save it to a secure location.

Step 2: Get the Packages Installed and the Libraries Ready
For R users:

Install the packages for developers, called `devtools`, provided by R. For accessing X data, we use the library `rtweet`. When installing these packages, if a warning message appears requesting that you download Rtools, you can download it and then install the package from https://cran.r-project.org/bin/windows/Rtools/. This is necessary only for Windows operating systems (OS), as macOS does not require it.

For Python users:

The package for accessing Twitter data is `tweepy`, which can be installed using `pip install tweepy`

Step 3: Write Code to Access the Tweets Data and Store the Data for Analysis

To write the code, you must first authenticate using your `bearer_token`. Ensure that the token is valid and properly configured.

Once authenticated, you can retrieve data based on your query, either by specifying search terms or by targeting specific user accounts on X.com.

After downloading the data, store it locally and convert it into a structured format such as a DataFrame. This allows you to apply various data analysis tools and techniques to process, visualize, and derive insights from the collected information.

Below is a sample R code:

```
Library(rtweet)

# setup token, replace those "AAA" with your own
Bearer_token <- "AAAAAAAAAAAAAAAAAAAAAAMED2"
auth <- rtweet::rtweet_app(bearer_token)

# specify the search keywords
query <- "electric vehicles"

# Get the relevant tweets using the API
tweets <- rtweet::search_tweets2(
  q = query,
  n = 10,
  token = auth,
  include_rts = FALSE,
  lang = NULL,  # no limit on language
  parse = TRUE)

# check whether any tweets were returned
if(nrow(tweets) > 0) {

  # get the data: create time, author id, language, and the tweet text
  tweet_data <- tweets %>%
    select(
      created_at = created_at,
      author_id = user_id,
      lang = lang,
      text = text
    )

  # save to CSV file
  write.csv(tweet_data, "electric_vehicles_tweets.csv",
          row.names = FALSE, fileEncoding = "UTF-8")

  print("Tweets saved to 'electric_vehicles_tweets.csv'")
} else {
  print(" No tweets found for your query.")
}
```

The Python code that performs the same tasks as the R code above is available for download in the online supplementary materials.

Some Use Cases and Applications

Free and public data sources can be applied to a variety of marketing use cases:

- Customer Segmentation
  Overlaying web behavior with demographic profiles for targeting.
- Brand Sentiment
  To understand the general public's opinion and listen to social discussions about a particular brand, we can access these discussions and obtain the

data through APIs. The sentiment of these discussions, as well as much other information, can be analyzed by leveraging the capabilities provided by large language models. The data access is discussed in this chapter; the usage of large language model (LLM) will be detailed in Chap. 12.

- Product Reviews
  Some websites offer a platform for customers to provide product reviews. It is possible to obtain the review data from websites like Amazon or Yelp to extract feature sentiment and inform product design and improvement.
- Competitor Analysis
  Some competitors' services, such as competing restaurants, may also be listed on Yelp. Their customer review data can be accessed through the API or the public dataset. Understanding what customers like or dislike about the competition provides valuable insights into how to improve your own offerings.
- Trend Tracking
  Data from any single organization is not enough to identify the market trend, especially at the early stage. To plan effectively, it is feasible to track trends by combining relevant data from Google Trends and discussions reflected in tweets posted on X.com.

Free data sources provide a powerful foundation for conducting customer analytics and marketing research. By combining internal data with external APIs and public datasets, organizations can develop a comprehensive understanding of their audience and the broader market landscape.

Customer Data Protection and Regulations

In the era of data-driven decision-making, business and their marketing efforts have increasingly come to rely on vast amounts of customer data to personalize experiences, optimize operations, and improve business outcomes. However, collecting and analyzing such data introduces significant ethical, legal, and operational responsibilities. Businesses must not only consider how to extract value from customer data, but also how to respect and protect the privacy of the individuals behind the data points. This is especially critical for online retailers, whose operations inherently involve the continuous collection and processing of user information.

Why Data Protection Matters

Modern marketing analytics involves tracking detailed user interactions, such as which pages customers visit, which products they click on, what they add to their carts, how they respond to marketing emails, and more. When combined with demographic and behavioral profiling, this allows for deep insights into individual preferences. However, from the customer's perspective, this same level of data use can feel invasive or even manipulative if not managed transparently and ethically.

Moreover, security breaches or mishandled data can damage trust and lead to regulatory penalties. The long-term value of data-driven marketing can only be sustained if companies build a foundation of trust, transparency, and legal compliance in their data collection practices.

Case Example: An Online Retailer
Consider an online retailer selling consumer goods across several countries. To improve its marketing performance, the retailer collects a wide range of data: customers' names, emails, IP addresses, browsing behavior, purchase history, and clickstream data from digital ads and newsletters. It uses this data to build user segments, predict future purchases, and personalize web and email content.

At the same time, this retailer is now responsible for storing and processing sensitive customer information. If a customer is based in the European Union, the retailer must ensure compliance with the General Data Protection Regulation (GDPR). If the customer is from California, it must comply with the California Consumer Privacy Act (CCPA). In Brazil, the Law of General Personal Data (LGPD) applies. Each regulation has its own definition of what data is considered personal, what rights consumers have, and how businesses must behave in data protection and usage.

The retailer's ability to operate internationally and maintain customer trust hinges on its ability to comply with these global standards while continuing to innovate with data.

Key Regulatory Frameworks and Their Implications

1. General Data Protection Regulation (GDPR) from the European Union

   The GDPR, implemented in 2018, is one of the most comprehensive privacy regulations in the world. It applies not only to businesses based in the EU but also to any organization that processes the personal data of EU residents.

   Key requirements include:

   - Informed consent: Customers must actively opt-in to data collection, and consent must be specific and informed.
   - Right to access and deletion: Users can request a copy of their data or ask for it to be deleted ("right to be forgotten").
   - Data minimization: Only data necessary for the stated purpose should be collected.
   - Privacy by design: Systems should be built to protect data from the start, not as an afterthought.
   - Data breach notification: Companies must report breaches within 72 hours.

Example implication for online retailers: When collecting data through cookies or tracking tools, retailers must inform users explicitly and obtain opt-in consent before storing or using any data for marketing purposes.

2. California Consumer Privacy Act (CCPA)

   The CCPA, enacted in 2020, applies to businesses that collect personal information from California residents and meet certain thresholds (e.g., $25 million in annual revenue or data from 50,000 or more individuals).

   Key consumer rights include:

   - Right to know what personal data is being collected, and how it is used.
   - Right to delete personal data upon request.
   - Right to opt out of the sale of personal data.
   - Non-discrimination: Companies cannot penalize users for exercising their privacy rights.

   Example implication for online retailers: The business must include a "Do Not Sell My Personal Information" link on their website, especially if they use customer data for ad targeting or share it with third parties.

3. Other Emerging Global Regulations
   - Brazil (LGPD): The Brazilian General Data Protection Law closely mirrors the GDPR, with a focus on data subject rights and corporate accountability.
   - Canada (PIPEDA): The Personal Information Protection and Electronic Documents Act requires meaningful consent and transparency in how data is handled.
   - China (PIPL): China's Personal Information Protection Law introduces strict localization requirements and cross-border data transfer controls.

Online retailers operating in multiple countries must be prepared to customize their data handling procedures for each jurisdiction or adopt a global privacy framework that adheres to the most stringent standards applicable in each jurisdiction.

Balancing Innovation and Privacy

Companies often fear that strict privacy rules will limit their ability to innovate with data. However, the opposite is increasingly true: demonstrating privacy-consciousness enhances customer trust, which can lead to deeper engagement and better long-term outcomes.

A retailer that transparently explains how it uses customer data, provides applicable controls, and avoids intrusive or irrelevant personalization is more likely to be perceived as responsible and trustworthy. That brand perception itself is a competitive advantage.

In fact, ethical data practices can open up opportunities. For instance:

- Customers may be more willing to share preferences if they believe the brand will use them responsibly.
- Data quality improves when users knowingly and willingly provide information.
- Data security reduces the risk of costly breaches or lawsuits.

As online retailers increasingly rely on customer data for personalized marketing, loyalty programs, and performance analytics, they must remain vigilant about protecting this data. Navigating global privacy regulations, such as GDPR and CCPA, requires not only legal compliance but also operational diligence and a customer-centric approach.

Ultimately, protecting customer data is not just about avoiding penalties; it's about building trust, demonstrating transparency, and ensuring long-term sustainability in a digital-first economy. Marketing that respects privacy is not a contradiction; it is a necessity and a differentiator.

Summary

Before delving into data analytics, this chapter lays the foundations for data-driven decision-making, focusing on data collection and the resources and tools available beyond an organization's internal data systems. External data has the capacity to augment information and resources that are typically inaccessible via internal channels. While such data is invaluable, it is not without limitations. A notable constraint stems from the use of Application Programming Interfaces (APIs) for extracting data from other platforms, as some entities impose strict restrictions on such activities, citing concerns related to data privacy.

For both internal and external data, the protection of data privacy and the ethical utilization of data sources serve as critical guidelines for data usage and analytics. This chapter also provides basic information on relevant regulations, non-compliance with which can result in substantial penalties for businesses. For instance, in July 2021, the European Union (EU) imposed a fine of $887 million on Amazon, finding the company in violation of the General Data Protection Regulation (GDPR) in its advertising practices. Similarly, in September 2022, Meta was fined $402 million by the EU for Instagram (a Meta-owned platform) failing to handle children's data appropriately, a violation of GDPR requirements.

Homework

Follow the instructions from the chapter, scrape some data from Twitter, put it in a dataframe, and then store it in a CSV file. Feel free to choose to use either programming language, R or Python.

CHAPTER 4

Media Mix Modeling

Learning Objectives

1. Explain the concept of media mix modeling (MMM) and the differences between aggregate-level and individual-level modeling approaches.
2. Apply MMM techniques to evaluate the returns on marketing investment across multiple channels.
3. Develop and incorporate ad-stock variables to capture carryover effects in marketing data analysis.
4. Interpret parameter estimates derived from an MMM.

Introduction

Media mix modeling (MMM) is a classical yet extremely popular tool in marketing analytics and probably the most important one in building the foundation of data-driven marketing decisions. Media mix refers to the collection of marketing channels a company uses to market and promote its products and brands. Media mix modeling evaluates the relative impact of different marketing media channels on business outcomes by analyzing media spending data using appropriate statistical models (Fig. 4.1).

Why media mix modeling is important:

Supplementary Information The online version contains supplementary material available at https://doi.org/10.1007/978-3-032-11130-2_4.

© The Author(s), under exclusive license to Springer Nature Switzerland AG 2026
X. Dong, *Marketing Analytics and Data Science*,
https://doi.org/10.1007/978-3-032-11130-2_4

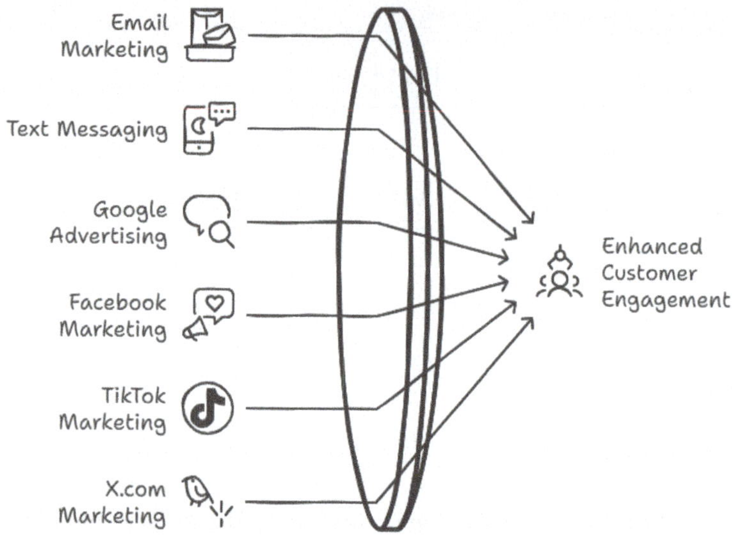

Fig. 4.1 Digital channels in media mix for enhancing customer engagement

Comprehensive Overview of Marketing Channels

Media mix modeling (MMM) offers a comprehensive view of all marketing channels employed by the marketing team. Keeping track of these channels can be more challenging than expected, especially for sophisticated marketing teams. Marketing is a dynamic field where tools and strategies evolve in response to changing business priorities and advancements in available technologies. Consistently monitoring these changes ensures continuous improvement and informed decision-making.

Integration and Standardization of Marketing Data Across Channels

MMM helps integrate data from various marketing channels, enabling meaningful and comparable analyses. Combining data from different sources can be complex. For example, a company may reach customers through both email and Facebook. While email marketing allows direct access to customer information, Facebook's data sharing is more restricted. Regulations like the General Data Protection Regulation (GDPR) further complicate data accessibility. Platforms such as Google and Facebook gain even more influence as a result of these privacy regulations. For a more detailed discussion, refer to the recommended reading Kostov, N., and Schechner, S. (2019) published on *The Wall Street Journal* (June 17, 2019).[1]

[1] Kostov, N., & Schechner, S. (2019). GDPR has been a boom for Google and Facebook. *Wall Street Journal.*

Foundation for Data-Driven Marketing Decisions

MMM builds the foundation for making marketing decisions based on systematic modeling and data analysis by integrating data across all marketing channels. While experience-based decision-making is still valuable, it lacks consistency and scalability. By leveraging data and modeling techniques, marketing teams can optimize resource allocation across different channels, ensuring efficient engagement with both potential and existing customers.

Classification of Media Mix Models

There are multiple ways to classify media mix models. One common approach is to categorize them based on the level of data aggregation:

1. **Aggregate-level analysis**: focuses on overall trends and patterns by examining data at a macro level, such as evaluating the effects of total ad spend and overall sales.
2. **Individual-level analysis**: examines data at a more granular level, such as customer-level behaviors and interactions, to analyze customer behaviors and marketing interactions to derive insights about marketing effectiveness.

Historically, these two approaches have risen and fallen in popularity at different times, largely reflecting changes in data availability and measurement technologies.

Before the digital age, collecting data at the individual customerlevel was very costly and often infeasible. As a result, most marketing, promotion, and sales data were primarily collected at the aggregate level.

Aggregate-level data does not track individual customers. Instead, they record marketing activities, such as advertising expenditures across traditional media channels, for different products, in different geographic locations, over time. Companies typically pair these data with sales data at the same product-market-time level. A common approach for analyzing the impact of these different promotional activities across media channels is to apply multiple regression model to these aggregate-level data.

In the digital age, promotional activities extend beyond traditional media. As customers spend more time online, firms increasingly rely on digital channels to reach their customers, such as email campigns, keyword advertising on Google, and ads on social media platfirms, such as Facebook. Unlike traditional offline activities, online interactions can be easily tracked at the individual level, including whether or when a user opens an email, clicks a link or interacts with an ad while browsing. This rich, behavior-level data has become a valuable resource for marketers, enabling much more detailed measurement and enhanced targeting.

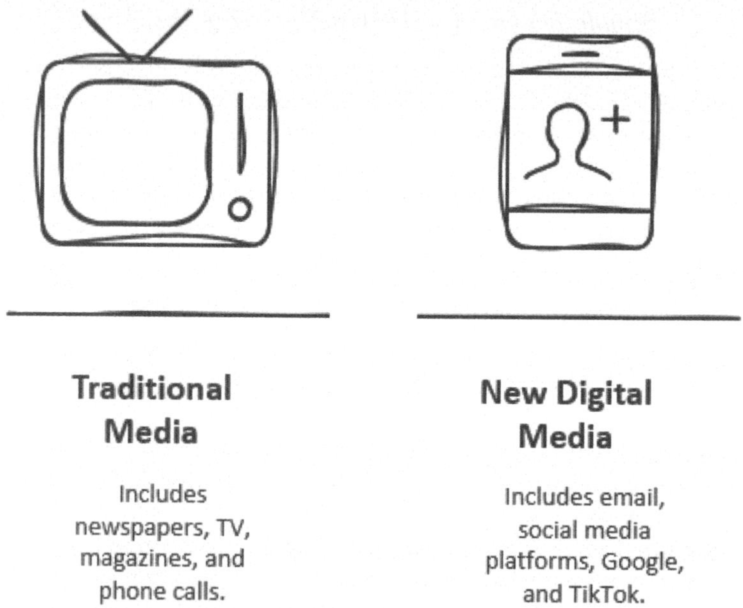

Fig. 4.2 Traditional versus digital media

In both aggregate and individual level analysis, the media channels may include all or some of the following channels, classified in two distinct groups:

1. Traditional media channels: TV, radio, newspaper, billboard on the road, etc.
2. Digital media channels: email, search ads, social media, mobile, etc. (Fig. 4.2)

Why digital marketing is attractive to marketers:

Cost-Effective

Digital marketing is significantly cheaper than traditional marketing media. For instance, airing a 30-second ad during the Super Bowl in 2025 cost more than $8 million, which was $7 million the year before. In contrast, running a similar video ad on YouTube may cost less than $100, and advertisers may only pay if viewers watch the ad rather than skipping it (Fig. 4.3).

Digital Ads Can Be Tracked

Although $8 million is spent on the Super Bowl, an advertiser can not identify which specific consumers actually saw the ads. This issue has created the classical frustration to marketers, as demonstrated in the famous quote by John

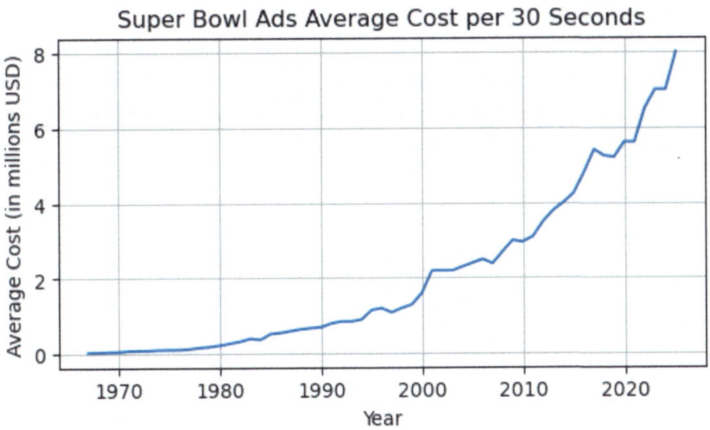

Fig. 4.3 Average Super Bowl ads cost for 30 seconds

Wanamaker, "I know half of my advertising dollars are wasted … I just don't know which half". In contrast, online ads can be tracked at the individual level. For example, if the Santa Clara University (SCU) marketing team bids on the keywords "MBA". Google can record which users saw the SCU MBA ad, which users clicked it, and when those actions took place.

General Process in Building a Media Mix Model

Before demonstrating MMM with an example, let me outline the general process of it.

Step 1: Understand the Problem and Gather the Necessary Data

The first step is to translate the business problem into a data analytics problem. Knowing that we are likely to solve this problem using a regression-type model, we need to find Y and X variables.

Define Y *Variable*

Identifying the right Y variable (dependent variable in the model) is not always straightforward. In practice, there may be multiple potential outcome variables to choose from. The key is to select an outcome variable that the marketing actions under evaluation can plausibly influence.

This often depends on the specific type of campaign being analyzed. In most cases, marketing campaigns can be categorized by their primary objectives, and the choice of the Y variable should be aligned with these objectives:

1. Short-term objective, such as

   - Instantaneous effect of marketing activities
   - Adjustments in marketing actions in response competition's marketing activities.

2. Long-term objective, such as

   - To have the brand better known to more people
   - To reposition the brand image or expand to a new product line

Evaluating long-term objectives requires a deeper understanding of the business context, such as the competitive landscape, industry trends, and future outlook. It also involves more advanced analytical techniques, including time-series analysis and dynamic modeling, to account for the influence of time on outcomes. These considerations are beyond the scope of this book.

In this book, we focus on the cases with short-term objectives.

To further define the problem, we need to consider two additional dimensions:

- First, whether the primary outcome of interest is the number of customers or the dollar value of product sales
- Second, the appropriate length of historical data to include in the analysis

The details of each consideration are discussed next.

Customer Versus Product-Focused Metrics

It is common to use product-focused metrics as the Y variable in marketing analytics, such as monthly revenue or units sold. However, these are not the only relevant metrics. The selection of outcome variables depends on the nature of the business. For example, in the high-tech sector, many companies operate as platforms that serve two-sided markets. In this case, the number of active users may be a more relevant outcome variable for the firm.

Consider LinkedIn: many users have accounts but do not pay for the service. LinkedIn's primary revenue comes from companies that pay to recruit talent through the platform. If user engagement declines, so does the platform's value to their paying clients. Therefore, maintaining and growing an active user base is critical to sustaining revenue for LinkedIn.

In such cases, customer-focused metrics may be more meaningful indicators of marketing effectiveness. Examples include:

- Number of customers
- Number of new customers acquired
- Number of daily or monthly active users (DAU or MAU)

These metrics reflect the health of the platform's user base, which is often the key driver of long-term business success in two-sided markets.

Time Dimension

Once the outcome metric is defined, the next step is to determine the appropriate time dimension to be included in the data, which includes:

- The granularity of the data
- The length of the historical period to include in the analysis

In online marketing, customer actions, such as webpage browsing and digital ad interactions, are often tracked at a very fine-grained level (e.g., minutes or seconds). However, marketing expenditure data, such as ad purchase decisions, may be recorded at different time intervals depending on the channel and data source. For example, spending on online channels may be captured daily, whereas spending on offline media, such as television or print, may be reported on a weekly or monthly basis. These marketing expenditure data are often included in the regression model as the X variables, while the customer actions are the Y variables. To conduct a regression analysis, these two sets of variables must be aligned. It is necessary to aggregate the data to a standard time scale so that the time dimensions are consistent across all variables. This may require rolling up customer behavior data or disaggregating marketing spend, depending on the available data.

Another important consideration is the length of historical data included in the analysis. Including a more extended time window (e.g., the past 10 years) can increase statistical power by providing more observations and a larger sample size. However, this comes with the risk that older data may no longer reflect current market dynamics. Business models, customer preferences, and competitive landscapes frequently undergo significant changes over time. Insights derived from outdated data may not be relevant for today's marketing decisions, especially in rapidly changing industries.

This highlights the need to strike a balance between statistical rigor (which benefits from more data) and business relevance (which requires current, contextually appropriate information). Making these decisions effectively often involves collaboration with domain experts who possess in-depth knowledge of the company's operations and market environment.

Choosing X *Variables for the Media Mix Model*

Once the Y variable is selected and the appropriate time range is defined, the next step is to identify the X variables for building the media mix model (MMM). This involves defining campaign objectives to collect relevant information on marketing activities and expenditures across various channels, as well as any other control variables that may have influenced the Y variable.

- Define campaign objectives and select relevant X and Y variables
- Marketing campaigns often serve multiple objectives, such as acquiring new customers, launching new products, or building brand awareness. As each objective is different, the marketing actions to analyze (the X variables) and the outcome variable (the Y variable) to use should align with the specific objective. For example, if the goal of the MMM analysis is to assess customer acquisition efforts., the X variables should include media spending and activtiies specifically tied to acquisition campaigns, and the Y variable should be a direct acquisition metric, such as the number of new customers acquired.
- Include appropriate control variables
- In addition to marketing inputs, it is essential to include control variables that account for other factors influencing the outcome. These may include:
  - Day-of-week or weekend effects
  - Holiday periods
  - Seasonality patterns
  - One-time events or market disruptions

- Control variables help isolate the actual impact of marketing from external influences that might influence the outcome variable and otherwise bias the results.
- Incorporate customer-level data (if available)
- If customer-level data is accessible, it can enhance the granularity and precision of the model. Relevant information may include:
  - Demographic or firmographic attributes, such as geographic location, age, gender for individual users (for business-to-consumers market), or company size (for buesiness-to-business market)
  - Customer history with the brand, such as tenure as a customer, cumulative purchases, engagement levels, or segmentation status

Step 2: Develop and Test the Model

This step heavily relies on our knowledge of the regression tool and its statistical features. The details are presented later through an example; the summary and quick review of the statistics are provided here:

First, choosing the transformation of Y, if needed

- Usually, it is either linear Y or its ln-transformation $\ln(Y)$.
- If the data records the dependent variable over a time period, in the format of time-series data, it may be necessary to decide whether to take the value of Y or the increments from the previous time period: $Y_t - Y_{t-1}$.

Second, choosing the transformations of *X*, if needed

- Create nonlinear transformations of some *X* variables
- Create dummy variables for some categorical variables in the *X*
- Create interaction variables between some demographic variables and some marketing variables
- Create stock variables to capture not only the instantaneous impact but also the lasting effect of the marketing spent in the past.

Plotting the data may help you determine whether any transformation is needed, and if so, which one to choose.

Step 3: Interpret the Results and Provide Suggestions

A commonly used metric in analyzing the results of marketing mix modeling (MMM) is Return on Advertising Spend (ROAS), which is conceptually similar to Return on Investment (ROI). ROAS measures the incremental return generated for each additional dollar spent on a specific advertising or marketing channel.

In a linear regression framework, where the *X* variables represent dollar spend for each channel and the *Y* variable is the outcome of interest (e.g., sales or customer acquisition), the regression coefficient for each channel directly represents its ROAS. That is, it estimates the marginal return in terms of the *Y* variable resulting from one additional dollar spend in that channel.

It is equally important to interpret the standard error associated with each coefficient, which reflects the uncertainty of the parameter estimate. While standard errors and confidence intervals are standard outputs in statistical modeling, they can be challenging to interpret.

It is up to the analyst to thoughtfully translate this uncertainty into a format that is accessible to business decision-makers. For example:

- When uncertainty is low (as is often the case with large datasets), it may be sufficient to focus on the point estimates of the model parameters and mention uncertainty only when specifically asked.
- When uncertainty is high, it becomes critical to communicate it effectively. Instead of reporting raw statistical metrics, the analyst might say: *"We are 90% confident that investing in Google Ads yields a higher return than TV advertising, based on the estimated model parameters and ROAS"*.

This type of interpretation allows stakeholders to make informed decisions while incorporating both the estimated effectiveness and the reliability of the results.

An Example of a Media Mix Model Using Aggregate-Level Data

When online marketing first rose to prominence in the early 2000s, many believed it would eventually replace traditional mass media. Over two decades later, however, traditional channels, such as television, newspapers, and radio, continue to command substantial marketing budgets, particularly for established consumer goods brands. While spending on these channels has gradually declined over time, they remain a significant component of the modern marketing mix.

In this section, as an example, we demonstrate how to apply marketing mix modeling (MMM) to traditional media channels, utilizing linear regression models to assess their effectiveness and return on investment. This book provides the R code needed for implementing this analysis. The equivalent Python code is available online as supplementary material.

First, Understand the Data

The data is from an automobile company, containing its monthly car sales from January 2012 to December 2015, along with their monthly ad spend across multiple media channels during the same time period. For privacy concerns, the company's name cannot be disclosed, and the data are also anonymized and rescaled.

```
cardf = read.csv("sales_ads_use.csv", header=TRUE, sep=",")
```

To set up a linear regression model, we need to find the Y variable and the X variables. In this case, the Y variable is the number of cars sold in each month.

Before modeling, plot the data and thoroughly understand the data-generating process.

```
yr = cardf$year
mn = cardf$month
yrmn = yr*100+mn
yrmndt =paste(as.character(yrmn),"01",sep="")
x = as.Date(yrmndt,"%Y%m%d")

plot(x,cardf$sales,type="l",ylim=c(0,3000))
```

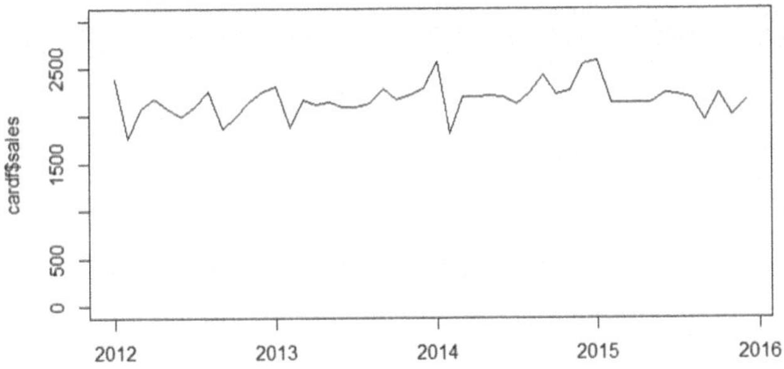

Second, Try Some Models

A common starting point in building a model is to include all available variables in the dataset as potential X variables. While this approach is convenient, it is essential to recognize a key limitation: in practice, we rarely have access to *all* the data we would like to have.

Therefore, it is essential to be aware of what information is missing and to thoughtfully consider the potential impact of these omissions on the modeling results. Understanding the limitations of your data and how they may bias or constrain your analysis is critical in building robust and interpretable models.

```
head(cardf)

##   year month sales newspaper magazine radio  tv
## 1 2012     1  2381        20      504    24 265
## 2 2012     2  1765       113       17    51   0
## 3 2012     3  2072        49      125   266   0
## 4 2012     4  2183       482      170    63   0
## 5 2012     5  2085       111      122   265   0
## 6 2012     6  1990       262       49    10   0
```

Let's start by adding all the channel expenditures for each channel in our data.

$$\text{Sales}_t = \beta_0 + \beta_1 \times \text{Newspaper}_t + \beta_2 \times \text{Magazine}_t + \beta_3 \times \text{Radio}_t + \beta_4 \times \text{TV}_t + \epsilon_t$$

In this model setup, we assume that marketing dollars will influence sales, and each channel has a linear impact on sales.

Let's try to estimate this model, and then see how we can improve it.

```
Xvars = as.matrix(cardf[,4:7])
Yval = cardf[,3]
lm_basic = lm(Yval~Xvars)
summary(lm_basic)

##
## Call:
## lm(formula = Yval ~ Xvars)
##
## Residuals:
##      Min       1Q  Median       3Q      Max
## -287.68  -52.80    1.75   41.25  327.53
##
## Coefficients:
##                 Estimate Std. Error t value Pr(>|t|)
## (Intercept)    1.942e+03  3.294e+01  58.953  < 2e-16 ***
## Xvarsnewspaper 3.721e-01  1.144e-01   3.254  0.00222 **
## Xvarsmagazine  4.023e-02  8.156e-02   0.493  0.62439
## Xvarsradio     2.918e-01  4.746e-02   6.149 2.22e-07 ***
## Xvarstv        2.894e-01  6.194e-02   4.673 2.93e-05 ***
## ---
## Signif. codes:  0 '***' 0.001 '**' 0.01 '*' 0.05 '.' 0.1 ' ' 1
##
## Residual standard error: 109.4 on 43 degrees of freedom
## Multiple R-squared:  0.6029, Adjusted R-squared:  0.5659
## F-statistic: 16.32 on 4 and 43 DF,  p-value: 3.329e-08
```

Understand the Regression Results

The printout from the R code contains the regression model's estimation results. To evaluate regression model estimates, we typically focus on two components.

First is the **coefficient** table. In the output shown above, this table has five columns. The first lists the names of the X variables. The second, **Estimate**, reports the estimated value of each model parameter. The third, **Std. Error,** reflects the uncertainty around that estimate. One useful way to think about this uncertainty metric is to treat each coefficient estimate as a random variable, following an approximately normal distribution, centered at the estimated value and with spread given by the standard error. Smaller standard errors indicate tighter distributions and therefore greater certainty about the estimate; larger standard errors imply more uncertainty. One important feature of standard error is that its value is directly related to the sample size. When the sample size is small, the model estimation results tend to be less certain, and hence the standard error values tend to be large, and vice versa. When the sample size is really large, without calculating it, it is know that the standard error values can be really small. Given the calculations of standard errors are much more costly than getting the parameter estimates, it is common that model results do not report standard errors when using a large sample size for a small number of parameters, for example, millions of data points for a few parameters.

In regression models, we are often most concerned with whether each coefficient is truly positive or negative, rather than simply whether the printed estimate is positive or negative. Because of estimation uncertainty, an estimate that appears positive or negative could still be consistent with a zero effect. A common benchmark is whether we can say we are at least 95% confident that the coefficient is above or below zero.

To quantify this, we examine the last two columns: t values and p-values. The **t-value** is computed as the ratio of the estimate to its standard error (Estimate / Std. Error). A larger absolute t-value means the estimate is large relative to its uncertainty, implying greater confidence about its sign. Under the normal approximation, a t-value greater than about 1.96 suggests roughly 95% confidence that the effect is positive; a t-value less than about –1.96 suggests roughly 95% confidence that the effect is negative. Many software outputs also summarize this information with significance markers (stars or dots) next to each coefficient, as shown in the R output above.

The same conclusion can be drawn from the **p-value**, which is computed under the null hypothesis that the true coefficient is zero (i.e., the corresponding X variable has no impact on Y). The p-value represents how likely it would be to observe an estimate at least as extreme as the one we obtained if the true effect were actually zero. A large p-value indicates that the data are consistent with the null hypothesis (i.e., no effect), while a small p-value suggests that the null hypothesis is unlikely and supports the conclusion that the variable has a statistically significant positive or negative association with Y.

The coefficient table evaluates the model estimates for each parameter. The second component assesses the model's overall fit to the data.

Second is the **R-squared values**, including Multiple R-squared and Adjusted R-squared. R-squared measures how much variation in Y the model explains: higher values indicate a better fit, with 1 representing a perfect fit (which almost never occurs in marketing settings). R-squared is useful for comparing models estimated on the same dataset, but it tends to increase as more predictors (X variables) are added, even if those predictors contribute little real explanatory power. For that reason, Adjusted R-squared is often preferred when comparing models with different numbers of parameters, because it penalizes unnecessary complexity.

These statistics are helpful, but they provide only one side of the evaluation. Ultimately, determining whether the regression specification is appropriate and what actions to take based on the estimates also requires domain knowledge about the data-generating process and business context, including how marketing actually influences outcomes and whether the model's assumptions are reasonable.

Building a regression model can feel simple and straightforward, especially with tools like R and Python. However, any statistical model comes with assumptions, and fitting the model means imposing those assumptions on the data. For example, in the above model specification, both the outcome variable Y and the predictors X have a subscript t, indicating a time-series data setting.

Using the same t for Y_t and X_t implies that marketing spending in period t affects the outcome within that same period. Before proceeding, it is important to ask whether this assumption is reasonable for the current problem, particularly in contexts such as automobile sales, where purchase decisions may respond to marketing with a delay.

To address this issue, several modeling strategies can be employed to capture the lagged or delayed effects of marketing activities.

Lagged X *Variables*

According to marketing theory, advertising should be viewed as an investment that delivers value not only in the short term but also over the long run. This is especially true for traditional advertising channels, where the impact of marketing efforts and expenditures is often not immediate.

To account for the lingering or delayed effects of advertising, we can extend the model by including lagged X variables, which refer to past values of marketing spend. This approach enables us to capture how the influence of a marketing campaign manifests in the future.

For example,

$$\text{Sales}_t = \beta_0 + \beta_1 \times \text{Newspaper}_{t-1} + \beta_2 \times \text{Magazine}_{t-1} + \beta_3 \times \text{Radio}_{t-1} \\ + \beta_4 \times \text{TV}_{t-1} + \epsilon_t$$

In this model specification, all the X variables have a subscript t-1, while the Y variable has a subscript t, indicating that the previous time period's marketing spending impacts the current time sales. To estimate this model, we need to process the data first so that in the same row for Y_t, we have the variables from the last time period X_{t-1}. As a result, the X variables span the time period from time 1 to $N-1$. The Y variable will take the values from time period 2 to N.

The following picture demonstrates the process of changing the column of the variable from Newspaper_t to a column Newspaper_{t-1}, and also to a column of Newspaper_{t-2} (Fig. 4.4).

In this example,

- To create the new column of Newspaper_{t-1}, we move the column Newspaper_t to **one row** below, then the **first row** in Newspaper_{t-1} is a missing value. As a result, we remove the first row of the whole data before modeling.
- To create the new column of Newspaper_{t-2}, we move the column Newspaper_t to **two rows** below, then the **first two rows** in Newspaper_{t-2} are missing values. As a result, we remove the first two rows of the whole data before modeling.

To compare the data before and after taking the lagged variables, next, we list those before taking the lags:

year	month	sales	newspaper	n_lag1	n_lag2
2012	1	2381	20	NA	NA
2012	2	1765	113	20	NA
2012	3	2072	49	113	20
2012	4	2183	482	49	113
2012	5	2085	111	482	49
2012	6	1990	262	111	482
2012	7	2094	45	262	111
2012	8	2254	76	45	262
2012	9	1873	177	76	45
2012	10	1997	4	177	76
2012	11	2150	16	4	177
2012	12	2257	381		4

Fig. 4.4 Illustration of creating the lagged variables

```
Y_X<- cbind(Yval,Xvars)
head(Y_X)

##      Yval newspaper magazine radio  tv
## [1,] 2381        20      504    24 265
## [2,] 1765       113       17    51   0
## [3,] 2072        49      125   266   0
## [4,] 2183       482      170    63   0
## [5,] 2085       111      122   265   0
## [6,] 1990       262       49    10   0
```

The R code below demonstrates how to calculate the lags for the X variables and adjust the Y variable accordingly. It involves two simple steps:

- Select X variables from period 1 to $N-1$, and remove the data from the **final** period.

Select the Y *variable from period 2 to* N, and remove the data from the **first** period.

```
N=length(Yval)
Xvarlag = Xvars[1:(N-1),] #X variables are from 1:N-1, lost row N
Yvalnew=Yval[2:N]   # Y variable takes the values from 2:N, lost row 1.
Y_Xlag <-cbind(Yvalnew,Xvarlag)
head(Y_Xlag)
```

```
##      Yvalnew newspaper magazine radio  tv
## [1,]    1765        20      504    24 265
## [2,]    2072       113       17    51   0
## [3,]    2183        49      125   266   0
## [4,]    2085       482      170    63   0
## [5,]    1990       111      122   265   0
## [6,]    2094       262       49    10   0
```

Comparing these two tables above, it is clear that the lagged variable is just a shift of all the X values from the original table.
A regression model can be estimated using the following R code.

```
lm_lagX=lm(Yvalnew~Xvarlag)
summary(lm_lagX)
```

```
##
## Call:
## lm(formula = Yvalnew ~ Xvarlag)
##
## Residuals:
##     Min      1Q  Median      3Q     Max
## -334.30  -81.95    5.48   66.02  432.86
##
## Coefficients:
##                     Estimate Std. Error t value Pr(>|t|)
## (Intercept)       2162.19742   47.35276  45.661   <2e-16 ***
## Xvarlagnewspaper     0.07780    0.16243   0.479   0.6344
## Xvarlagmagazine     -0.20677    0.11630  -1.778   0.0827 .
## Xvarlagradio         0.14045    0.06786   2.070   0.0447 *
## Xvarlagtv           -0.14000    0.08791  -1.593   0.1188
## ---
## Signif. codes:  0 '***' 0.001 '**' 0.01 '*' 0.05 '.' 0.1 ' ' 1
##
## Residual standard error: 155.3 on 42 degrees of freedom
## Multiple R-squared:  0.1899, Adjusted R-squared:  0.1128
## F-statistic: 2.462 on 4 and 42 DF,  p-value: 0.05986
```

Lagged Y *Variables*

If you believe some continuance of car sales exists over time, you can consider adding a lagged Y variable as an additional predictor variable.

$$\begin{aligned}\text{Sales}_t &= \beta_0 + \beta_1 \times \text{Newspaper}_{t-1} + \beta_2 \times \text{Magazine}_{t-1} + \beta_3 \times \text{Radio}_{t-1} \\ &\quad + \beta_4 \times \text{TV}_{t-1} + \beta_5 \times \text{Sales}_{t-1} + \epsilon_t\end{aligned}$$

To do that, we need to create the lagged Y variable first, before adding it to the predictors in the regression model.

```
lagY=Yval[1:(N-1)]
lm_lagY = lm(Yvalnew~Xvarlag+lagY)
summary(lm_lagY)

##
## Call:
## lm(formula = Yvalnew ~ Xvarlag + lagY)
##
## Residuals:
##      Min       1Q   Median       3Q      Max
## -337.10   -83.17     5.77    65.82   433.62
##
## Coefficients:
##                    Estimate Std. Error t value Pr(>|t|)
## (Intercept)       2.181e+03  4.295e+02   5.077 8.74e-06 ***
## Xvarlagnewspaper  8.135e-02  1.834e-01   0.444   0.6597
## Xvarlagmagazine  -2.063e-01  1.181e-01  -1.747   0.0882 .
## Xvarlagradio      1.433e-01  9.464e-02   1.514   0.1377
## Xvarlagtv        -1.372e-01  1.095e-01  -1.253   0.2174
## lagY             -9.608e-03  2.202e-01  -0.044   0.9654
## ---
## Signif. codes:  0 '***' 0.001 '**' 0.01 '*' 0.05 '.' 0.1 ' ' 1
##
## Residual standard error: 157.2 on 41 degrees of freedom
## Multiple R-squared:   0.19,  Adjusted R-squared:  0.09116
## F-statistic: 1.923 on 5 and 41 DF,  p-value: 0.1114
```

However, given that the parameter for lag$Y$ is not statistically significant, it does not significantly improve the model fit.

Nonlinear Transformation of the X *Variables*

In the regression model above, all X variables are included in a linear form, implying a constant marginal effect: that is, each additional unit of marketing spend is constrained to have the same impact on the outcome (Y) value, regardless of the current level of ad spending in the channel.

However, in practice, marketing often exhibits diminishing returns, which shows that after a certain level, each additional dollar yields progressively smaller gains. To capture this pattern, we can apply nonlinear transformations to the X variables before including them in the regression model.

A nonlinear transformation involves applying a mathematical function to the original variable to better reflect the underlying relationship. Two of the most commonly used approaches are:

- Adding a squared term of the X variable, in addition to the linear X, to the model specification
- Replacing the linear X with a ln-transformation of the X variable, $\ln(X)$

When adding a square term, we also keep the linear term in the model. To do that, we can use the following code:

```
Xvarlag2 = Xvarlag^2/1000
lm_sqr= lm(Yvalnew~Xvarlag+Xvarlag2+lagY)
summary(lm_sqr)

##
## Call:
## lm(formula = Yvalnew ~ Xvarlag + Xvarlag2 + lagY)
##
## Residuals:
##      Min       1Q   Median       3Q      Max
## -281.85   -78.20    -7.07    50.43   403.04
##
## Coefficients:
##                      Estimate Std. Error t value Pr(>|t|)
## (Intercept)         2574.5643   377.4178   6.822 4.91e-08 ***
## Xvarlagnewspaper      -0.5659     0.4232  -1.337 0.189293
## Xvarlagmagazine       -0.4397     0.2410  -1.825 0.076111 .
## Xvarlagradio           1.0374     0.2214   4.686 3.70e-05 ***
## Xvarlagtv              0.2259     0.2169   1.042 0.304256
## Xvarlag2newspaper      2.1452     1.0598   2.024 0.050208 .
## Xvarlag2magazine       0.4517     0.2914   1.550 0.129639
## Xvarlag2radio         -0.6921     0.1698  -4.075 0.000234 ***
## Xvarlag2tv            -0.3006     0.2271  -1.324 0.193730
## lagY                  -0.2848     0.1974  -1.443 0.157469
## ---
## Signif. codes:  0 '***' 0.001 '**' 0.01 '*' 0.05 '.' 0.1 ' ' 1
##
## Residual standard error: 133.3 on 37 degrees of freedom
## Multiple R-squared:  0.4741, Adjusted R-squared:  0.3462
## F-statistic: 3.707 on 9 and 37 DF,  p-value: 0.002155
```

When the estimated coefficient on the squared term is negative, it suggests diminishing returns, which indicates that each additional dollar spent yields a smaller incremental impact. However, it also implies that beyond a certain level of spending, additional investment could actually reduce overall sales. This can be a difficult message to convey to marketing teams, who may find it counter-intuitive or implausible.

The figure below illustrates this point. As shown in the figure, when the X value is below 1.5, the curve reflects diminishing returns, which is consistent with what we aim to capture. However, once X exceeds 1.5, the curve slopes downward, indicating that additional spending could negatively impact the outcome. This implication may be problematic from a business perspective and is often not well-received by decision-makers (Fig. 4.5).

To avoid the undesired portion of the decreasing returns, we can use the ln-transformation. However, the ln-transformation requires all values to be positive, as by definition $\ln(0) = -\infty$. To avoid taking the ln of a value 0, as sometimes X variables may have 0 values, we use $\ln(x + 1)$, as long as all x's take non-negative values.

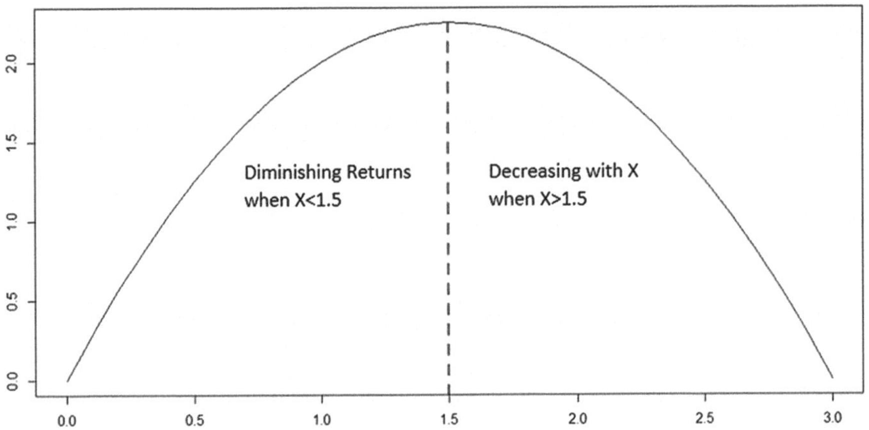

Fig. 4.5 Illustration for a square transformation

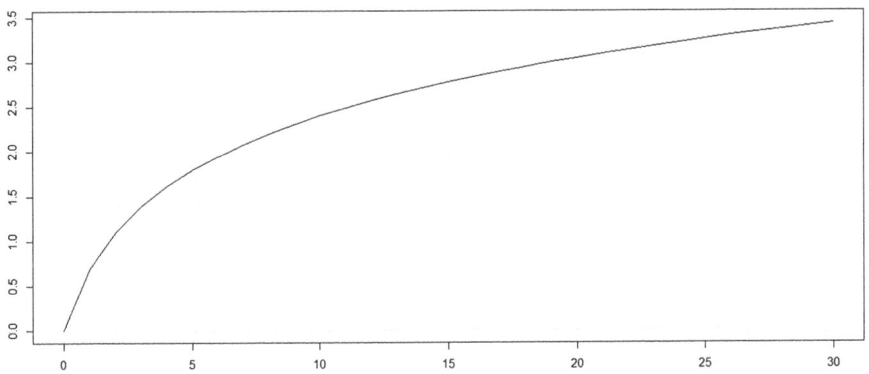

Fig. 4.6 Illustration for an ln-transformation

The ln-transformation can be illustrated using the plot in Fig. 4.6.

Next, we estimate the model using the ln-transformation of marketing spending across all channels. In this case, we no longer need to include the linear X variables.

```
lnX = log(Xvarlag+1)
lm_lnx = lm(Yvalnew~lnX+lagY)
summary(lm_lnx)
```

```
##
## Call:
## lm(formula = Yvalnew ~ lnX + lagY)
##
## Residuals:
##     Min      1Q  Median      3Q     Max
## -296.76  -68.83  -16.84   73.55  435.44
##
## Coefficients:
##                Estimate Std. Error t value Pr(>|t|)
## (Intercept)  2082.85289  328.40747   6.342 1.41e-07 ***
## lnXnewspaper   -1.75846   10.92381  -0.161   0.8729
## lnXmagazine   -11.16553   10.92088  -1.022   0.3126
## lnXradio       41.14276   20.02651   2.054   0.0463 *
## lnXtv         -15.18202   10.15150  -1.496   0.1424
## lagY           -0.01792    0.17851  -0.100   0.9205
## ---
## Signif. codes:  0 '***' 0.001 '**' 0.01 '*' 0.05 '.' 0.1 ' ' 1
##
## Residual standard error: 160.1 on 41 degrees of freedom
## Multiple R-squared:  0.1597, Adjusted R-squared:  0.0572
## F-statistic: 1.558 on 5 and 41 DF,  p-value: 0.1934
```

Adding a square-term of the X variable or taking a ln-transfomration are two most common approaches to incorporating non-linear transformation into a marketing analytics model. These mathematical operations have their own properties, whether it fits the model for marketing analytics is a judgement call that requires analyst's understanding of the marketing practice. For example, if the marketing expenditures never exceeds 1.5, as illustrated in the above figure, it is reasonable to add the squared term, as we do not need to worry about the negative slope region.

Ad-Stock Variables

So far, we have only plotted the *Y* variable, not any of the *X* variables yet. As an example, let's plot the newspaper spending data.

```
x = as.Date(yrmndt,"%Y%m%d")
plot(x,cardf$newspaper,type="l",ylim=c(0,600))
```

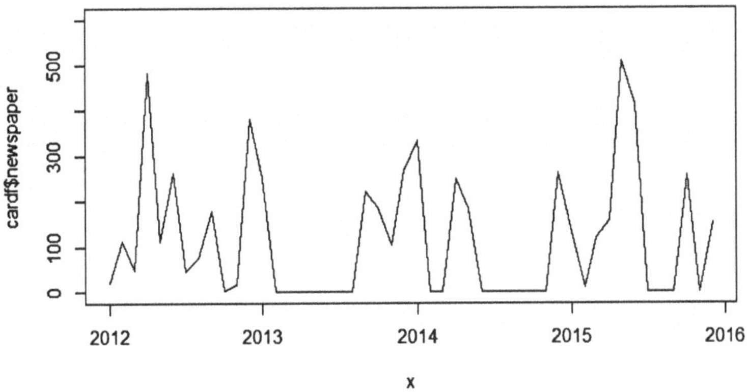

Pulsing in Advertising and Its Implications for Modeling

The spending pattern shown in the figure illustrates a high degree of variability. In some months, there is no advertising spend, which is often followed by periods of intense investment. This pattern is known in marketing as "pulsing".

Pulsing refers to a strategy where advertisers concentrate a large portion of their budget into short bursts of activity, followed by periods with little or no spending. This approach is widely used because marketers have found that concentrated exposure to ads can quickly increase consumer awareness and product knowledge. Importantly, this awareness does not vanish immediately when advertising stops. Instead, it tends to persist for a period of time. Before the effects fully decay, another pulse of spending is often introduced. This cyclic strategy has been proven to be both effective and cost-efficient and is therefore commonly used in both traditional and digital advertising.

In one of my previously published research projects (Shi & Dong, 2011),[2] we examined the pulsing strategy in the context of online advertising. Specifically, in keyword bidding environments, we found that pulsing leads to better performance than continuous spending. This reinforces the broader understanding that pulsing is beneficial across both traditional and digital media channels. As a result, we often find pulsing patterns in ad spending data across various marketing campaigns.

However, this strategy poses challenges for statistical modeling, particularly for regression analysis. Traditional regression assumes that independent variables (X) are random or naturally occurring, whereas pulsing is a deliberately engineered spending pattern. This mismatch can lead to biased or misleading estimates if not adequately addressed.

To account for this, it's essential to incorporate the concept of lagged effects, which is the fundamental assumption underlying the use of the pulsing strategy.

[2] Shi, Savannah Wei and Dong, Xiaojing, "The Effects of Bid-Pulsing on Keyword Performance in Search Engines", *International Journal of Electronic Commerce*, 2014 19(2), pages 3–38.

For example, if $100 is spent on advertising today, the effect is assumed not to be confined to today's sales alone. That same $100 may still influence tomorrow's sales, however, to a lesser degree. Similarly, advertising spend from two days ago might still have a small residual effect today. This concept of time-decayed impact underlies both the pulsing strategy and the need for appropriate modeling techniques, such as lagged variables or advertising stock transformations, to capture the carryover effect of advertising spend.

We can use the **ad-stock variable,** which is calculated as a time-decayed cumulative sum of all past advertising spend, capturing the lingering effect of prior marketing efforts.

It can be calculated as:

$$G_t = M_t + \lambda G_{t-1}$$

- G_t represents the ad-stock variable value at t, capturing all the ad spending dollars with time-related decay from previous spending.
- G_{t-1} represents the ad-stock value in the previous time period $t-1$.
- M_t represents the ad spending in the time period t. M_t is the same as X_t in the above regression.
- λ captures the discount due to time-related decay.

To understand this variable, suppose $t = 3$,

$$\begin{aligned} G_3 &= M_3 + \lambda G_2 \\ &= M_3 + \lambda\left(M_2 + \lambda G_1\right) = M_3 + \lambda M_2 + \lambda^2 G_1 \\ &= M_3 + \lambda M_2 + \lambda^2\left(M_1 + \lambda G_0\right) = M_3 + \lambda M_2 + \lambda^2 M_1 \end{aligned}$$

where $G_0 = 0$.

It demonstrates that the ad-stock variable at $t = 3$ captures

- The current period's ad spending
- A discounted ad spending of $t-1$ with discount λ
- A discounted ad spending of $t-2$ with discount λ^2

To implement it, the following R function is created, and can be called when creating the ad-stock variable based on the original per-period marketing spending variables stored in M_t.

```
# AdStockMatrix() Returns the scale matrix for creating the adstock variable
from the marketing per-period expenses
# Input:
#    lambda - the discounted value
#    MM - the matrix of the marketing variables

AdStockMatrix <- function(lambda, MM) {

  n=nrow(MM)

  r <- lambda^(seq_len(n)-1)
  m <- matrix(rep(r,n),nrow=n)

  z <- matrix(0,nrow=n,ncol=n)
  z[lower.tri(z,diag=TRUE)] <- m[row(m) <= (n+1-col(m))]
  GG <- z %*% MM
  GG

}
```

To create the AdStock G_t matrix from the current per-period marketing matrix M_t, all you need is to specify the M_t matrix, and then call the function, as illustrated in this script:

```
MM <- as.matrix(cardf[,4:7])

GG2 <- AdStockMatrix(0.2,MM)
GG8 <- AdStockMatrix(0.8,MM)
```

We created two series of the G_t matrix, with different values of λ, 0.2, and 0.8. To understand the implications of λ, we plot the two G_t values with different λ below.

```
a = cbind(MM[,1],GG2[,1],GG8[,1])
colorseq <- c("black","red","blue")
matplot(a,type="l",col=colorseq,lwd=2)
legend('top',c("original data","AdStock-0.2","AdStock-0.8"),lty=1:3,col=color
seq,cex=.75)
```

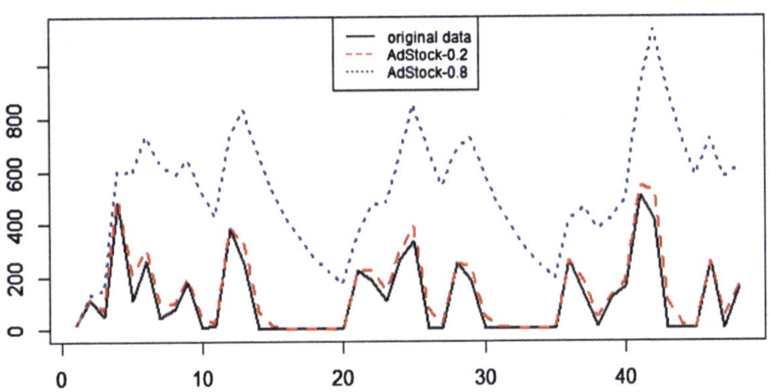

As illustrated in the R output, higher λ value (blue dotted line) indicates a more lingering impact of previous marketing influences than the lower value of λ (red dashed line). A λ close to 0 is very similar to using the original per-period marketing variable.

```
lm_adstock2 <- lm(Yval~GG2)
summary(lm_adstock2)

##
## Call:
## lm(formula = Yval ~ GG2)
##
## Residuals:
##      Min       1Q  Median      3Q     Max
## -272.82  -48.01   -4.81   54.28  379.13
##
## Coefficients:
##               Estimate Std. Error t value Pr(>|t|)
## (Intercept)  1.913e+03  4.054e+01  47.179  < 2e-16 ***
## GG2newspaper 3.592e-01  1.159e-01   3.100 0.003412 **
## GG2magazine  1.412e-02  8.373e-02   0.169 0.866920
## GG2radio     2.684e-01  4.510e-02   5.951 4.31e-07 ***
## GG2tv        2.577e-01  6.177e-02   4.172 0.000144 ***
## ---
## Signif. codes:  0 '***' 0.001 '**' 0.01 '*' 0.05 '.' 0.1 ' ' 1
##
## Residual standard error: 116 on 43 degrees of freedom
## Multiple R-squared:  0.5539, Adjusted R-squared:  0.5124
## F-statistic: 13.35 on 4 and 43 DF,  p-value: 3.743e-07
```

Finally, Results Interpretations

With the model estimates output in hand, our goal is to translate them into actionable insights for the marketing team, particularly by informing more effective budget allocation across these advertising channels.

In models that incorporate ad-stock variables, interpreting the parameter estimates becomes less straightforward than in models that use raw advertising spend. In this case, each parameter reflects the marginal impact of the ad-stock variable on the outcome (Υ), rather than the direct effect of spending.

The ad-stock variable captures the cumulative and decaying influence of past advertising. The marginal effect of spending an additional \$1 will depend on the timing of that investment. For simplicity, we can interpret the marginal effect as the addition of \$1 in spending in the current period, before any decay is applied, when interpreting the estimates for the ad-stock variables.

In this example, all parameter estimates are statistically significant. The results suggest that newspaper advertising has the highest marginal impact on the outcome, followed by TV and radio. In contrast, magazine advertising appears to be the least effective, indicating that it may be a good candidate for budget reduction.

Summary

This chapter presents one application of regression analysis in marketing, media mix modeling.

The goal of MMM is to evaluate the differential impact of marketing spend across channels, so we can adjust investment levels in each channel accordingly.

The data used in MMM:

- Spending data from each marketing media channel.
- Some business outcome variable.
- It is generally in the format of time series data.

It applies the regression approach:

- *Y* variable is the outcome of the marketing investment, such as sales, or the number of newly acquired customers, etc.
- *X* variables are the dollar spent in each channel. To understand the statistical features of the regression model estimates, we evaluate: \2022 Each parameter's estimate in the Coefficient table. The estimate with a large p-value (larger than 0.05 or 0.1) indicates the corresponding X variable has no statistically significant impact on the Y variable. \2022 The R-squared value, the model with a larger R-squared value fits the data better than the one with a lower R-squared value.

These variables can be included with nonlinear transformations, including:

- The squared term X^2, together with the linear term X. The issue is whether it falls into the decreasing region.
- The ln-transformation $\ln(X)$ or $\ln(X + 1)$ can also be utilized to capture diminishing returns.

Add lagged *Y* variables to capture the stability of the sales over time.

Add lagged *X* variables to allow previous spending in marketing to influence the current *Y* variable.

- We can create X_{t-1} and X_{t-2}, …, X_{t-k} to capture the lagged effect of X. However, we will lose some data because the first few data points do not include these lags.
- We can also create ad-stock variables to capture the common marketing practice of pulsing.

Homework

1. Explain media mix model:
   (a) What are the purposes of developing such an MMM?
   (b) What data are required to estimate an MMM?
   (c) Which statistical model can you use to build an MMM? Briefly describe it.
2. Make a time series plot for each channel spending, such as newspaper, magazine, etc. Comment on the pulsing behavior in terms of:
   (a) Why do companies want to employ pulsing strategy when allocating their advertising budget over time?
   (b) What statistical challenge does this lead to?
   (c) Explain in words (not in an equation) what an ad-stock variable is.
3. Write code and re-estimate the model with different lambda values, from 0.1, 0.2, 0.3, ..., 0.9. Compare the R-square values from each model and comment on the lambda value that yields the highest R-square value from the model estimation results.

CHAPTER 5

Market Segmentation and Clustering Analysis

Learning Objectives

1. Distinguish behavioral segmentation from traditional demographic-based segmentation approaches.
2. Explain the data required for conducting behavioral segmentation.
3. Apply K-means, hierarchical clustering, and DBSCAN methods for behavioral segmentation.
4. Compare the strengths and limitations of K-means, hierarchical clustering, and DBSCAN methods for behavioral segmentation.

Introduction

Market segmentation refers to the process of analyzing a company's total customer base, including business customers and individual consumers, and dividing it into distinct customer groups. It is a crucial function of the marketing department, playing a key role in achieving business success. Nearly all companies require a well-defined market segmentation strategy, as it builds the foundation for identifying the right customer groups. An effective market segmentation strategy enables companies to tailor their products, promotional messages, pricing, and marketing channels (the 4Ps) to reach their target audience with the desired impact.

The goal of market segmentation is to divide the large number of customers in a market into distinct subsets of customers, so that it achieves two criteria at the same time: the members are similar to one another within each segment

Supplementary Information The online version contains supplementary material available at https://doi.org/10.1007/978-3-032-11130-2_5.

© The Author(s), under exclusive license to Springer Nature Switzerland AG 2026

X. Dong, *Marketing Analytics and Data Science*,
https://doi.org/10.1007/978-3-032-11130-2_5

(**homogeneous within**) and different across segments (**heterogeneous across**).

How can this be achieved? In this chapter, we begin by reviewing traditional methods of market segmentation, which rely on the understanding of the business and its overall customer base while requiring limited data. We then introduce behavioral segmentation, a data-driven approach that groups customers based on their recorded behaviors, enabling more effective targeting strategies.

Common Ways to Segment the Market

Traditionally, market segmentation is based on *who they are*, using easily observable customer characteristics, such as demographics, geographics, or psychographics. It is common for companies to offer a range of products tailored to different customer groups, often defined by factors like age, income, or geographic location for consumer market, and firm size, industry or revenue for business market one popular segmentation tool based on consumer psychographics is the VALS survey, developed by Strategic Business Insights, a company located in Menlo Park, California.

The VALS survey asks a series of lifestyle questions, together with some demographic information such as age range, income range, and educational background. Based on the answers to these questions, each individual is assigned to one of the eight groups based on the following two dimensions (as shown in Fig. 5.1):

1. Vertical dimension: level of resources.
2. Horizontal dimension: lifestyle. Those on the right are more likely to try new fashions than those on the left (Fig. 5.1).

VALS enables marketers to understand consumers not only by who they are, but also by their interests and lifestyles. This approach provides more insights into customers before segmenting them. This approach offers a deeper understanding of consumers and is considered an improvement over segmentation based solely on observed demographics or geographic information. It has garnered considerable attention since its development in 1978. The question is, after over several decades, can we do better now?

The segmentation approaches mentioned above focus on the question of "who customers are". While this information is valuable, it alone is not sufficient to fully understand customer preferences and their likelihood of purchasing a company's products or services.

Behavioral segmentation takes a different approach by analyzing *what customers do*, including their actions, habits, and purchasing behaviors, and provides deeper insights beyond just demographic or psychographic traits.

Fig. 5.1 VALS framework

Behavioral Segmentation and Targeting: The General Process

Behavioral segmentation requires access to detailed data on individual customer actions and activities. In the past, such data was limited or unavailable, making this type of segmentation difficult to implement. However, with the rise of digital platforms, it has become increasingly common for companies to collect rich behavioral data, such as website browsing history, ad link clicks, product searches, and purchase records. These data provide valuable insights into each customer's activities and interactions with the website, which reflect their preferences and decision-making patterns.

With this behavioral tracking data, companies can apply data analytics techniques to perform market segmentation based on actual customer behavior.

The general process for behavioral segmentation typically involves the following three steps:

Step 1: Define the Goal

Before applying any analytical tools, the first and most crucial step in the segmentation process is to define the objective clearly and precisely. Without a well-defined goal, there is a significant risk of using inappropriate data and drawing misleading conclusions.

For example, if the goal is to establish a fundamental customer metric, such as categorizing customer loyalty levels, businesses often segment customers into tiers like Platinum or Gold. In these cases, total purchase value is typically the primary metric for customer segmentation, as seen in frequent flyer programs and hotel loyalty schemes.

In another scenario, a company may aim to forecast customer lifetime value (CLV) over an extended period, such as for the next ten years. Given the variability in customer needs and interactions, a single predictive model may not perform well across the entire customer base. In this case, segmentation can enhance predictive accuracy by training separate models tailored to distinct customer groups, which can be defined by metrics such as total past spending, purchase frequency, or tenure. In this particular scenario, segmentation is not the ultimate goal, rather, it facilitates customer behavior modeling and predictions. Despite the differences between the above two scenarios, both examples illustrate situations where segmentation is based on a single key metric, allowing for direct comparison across customers.

However, in many real-world applications, relevant customer behavior metrics are not so clearly defined. Instead, companies must rely on high-dimensional behavioral data, such as browsing history, ad interactions, and engagement patterns, to infer meaningful segments. In such cases, more advanced machine learning techniques are required to extract patterns and conduct effective segmentation.

The following sections will explore how to implement these techniques for behavior-based market segmentation using high-dimensional data.

Step 2: Conduct the Analysis

Clustering analysis helps identify groups of individuals who exhibit similar patterns across high-dimensional behavioral data, grouping those with similar activities while distinguishing them from others with different patterns.

The goal is to segment customers based on similarities in their behaviors, which are captured through multiple variables. The detailed steps of this analysis will be illustrated in a later section using an example with coding and data.

Step 3: Interpret the Results

Depending on the analysis goal, interpret the results and guide the next steps.

An Example of Behavioral Segmentation

Now, let's bring in a sample data set and execute behavior-based segmentation, applying two machine learning tools, clustering analysis and discriminant analysis.

Now, we get into R programming for the data analysis, the Python scripts and their outputs are included in the supplementary material online.

Before opening the file or conducting any R programming, a good practice is to get R ready by clearing prior memory and opening the packages we will need:

```
rm(list=ls())          # Remove anything that might currently be in memory so we
 can start fresh.
library(MASS)          # Load the MASS package
```

The `MASS` package is very useful in R for statistical analysis, so it is common practice to load these packages when conducting data analysis. MASS stands for "Modern Applied Statistics with S", as R was developed from the language S.

With the coding environment ready, read in the data, and give a name to the loaded data.

```
dataset <- read.csv('DurData_custpurchase.csv')
```

Take a quick peek at the data set.

```
class(dataset)  # it is a dataframe

## [1] "data.frame"

dim(dataset)    # the number of rows and columns in the data.frame

## [1] 19936    40

str(dataset)    # the basic structure of the dataset, including the first few
 numbers of each variable (column)

## 'data.frame':    19936 obs. of  40 variables:
##  $ X                            : int  1 2 3 4 5 6 7 8 9 10 ...
##  $ Household_ID                 : int  100003544 100012312 100016237 1000
22945 100022976 100024091 100024909 100025614 100025901 100026342 ...
##  $ N_catg1                      : int  0 0 0 0 1 0 1 0 3 0 ...
##  $ N_catg2                      : int  0 0 0 0 0 0 0 0 0 0 ...
##  $ N_catg3                      : int  0 0 0 0 0 0 0 0 0 0 ...
##  $ N_catg4                      : int  0 0 0 0 0 0 0 0 0 0 ...
##  $ N_catg5                      : int  0 0 0 1 0 0 0 0 0 2 ...
##  $ N_catg6                      : int  0 0 0 0 0 0 0 0 0 0 ...
##  $ N_catg7                      : int  0 0 0 0 0 4 2 0 2 0 ...
##  $ N_catg8                      : int  0 0 0 2 0 0 7 2 1 0 ...
##  $ N_catg9                      : int  0 0 0 0 0 0 1 0 0 0 ...
##  $ N_catg10                     : int  0 0 0 0 0 0 0 0 1 1 ...
##  $ N_catg11                     : int  0 1 1 6 0 0 3 0 2 6 ...
##  $ N_catg12                     : int  0 0 0 1 0 0 4 0 0 3 ...
##  $ N_catg13                     : int  0 1 0 0 0 2 3 0 9 0 ...
##  $ N_catg14                     : int  1 0 0 0 1 0 0 0 3 1 ...
##  $ N_catg15                     : int  0 0 0 0 0 0 1 0 0 0 ...
##  $ E_catg1                      : num  0 0 0 0 600 ...
##  $ E_catg2                      : num  0 0 0 0 0 0 0 0 0 0 ...
##  $ E_catg3                      : int  0 0 0 0 0 0 0 0 0 0 ...
##  $ E_catg4                      : num  0 0 0 0 0 0 0 0 0 0 ...
##  $ E_catg5                      : num  0 0 0 260 0 ...
##  $ E_catg6                      : num  0 0 0 0 0 0 0 0 0 0 ...
##  $ E_catg7                      : num  0 0 0 0 0 ...
##  $ E_catg8                      : num  0 0 0 25 0 ...
##  $ E_catg9                      : num  0 0 0 0 0 ...
##  $ E_catg10                     : num  0 0 0 0 0 ...
##  $ E_catg11                     : num  0 20 90 194 0 ...
##  $ E_catg12                     : num  0 0 0 150 0 ...
##  $ E_catg13                     : num  0 9.99 0 0 0 ...
##  $ E_catg14                     : num  100 0 0 0 250 ...
##  $ E_catg15                     : num  0 0 0 0 0 0 0 0 0 0 ...
##  $ Household_Income             : int  6 1 NA 5 7 5 7 NA 7 9 ...
##  $ Age_Household_Head           : int  28 24 NA 44 54 44 56 NA 72 48 ...
##  $ Number_Of_Kids               : int  0 0 0 1 0 0 0 0 0 3 ...
##  $ Total_Number_Of_Transactions : int  1 1 1 1 1 1 3 1 4 1 ...
##  $ Number_Of_Online_Purchase    : int  0 0 0 0 0 0 0 0 0 0 ...
##  $ Number_Of_Stores_Visited     : int  1 1 1 1 1 1 3 1 4 1 ...
##  $ Number_Of_Returns            : int  0 0 0 2 0 0 7 0 12 0 ...
##  $ Number_Of_Different_Categories: int  1 2 1 4 2 2 8 1 7 5 ...
```

- `class()` tells the class of the `dataset` is `data.frame`; therefore, we can use the functions designed for `data.frame`.

- `dim()` tells the dimensions of the `dataset`, with 19,936 rows and 40 columns.
- `str()` gives the structure of the data table, including all the information above; in addition, it lists the variable names and some data values for each variable. This helps us to get some sense of the data.

As with all data analysis, before building any models, it is necessary to answer a series of questions to obtain a comprehensive understanding of the data, such as

- What is the story behind the data? How were these data generated and how were they collected? This addresses the data generation process (DGP) of the data.
- Who or what does one row in the data represent? This directly speaks to the granularity of the data.
- How many observations are there? This indicates the data size in terms of the number of rows.
- What is the meaning of the variables (the columns)? This task may take some time to fully understand how the data in each column was measured and collected. For example, if the column is the age of each customer, it is necessary to know when the age data was collected. It could be collected when the customer first signed up for the service. If the data spans a long history, some customers may be the same age, but joined in different years. In that case, the data may record them as having different ages. It is therefore important to understand how the information is recorded.
- What type of data is in each variable? Numbers? Strings? Something else? When dealing with strings, we may need to convert them into either categorical variables or dummy variables before they can be included into our statistical models.

To answer the above questions and gain a thorough understanding of the data, additional context is often necessary. This information typically comes from supporting documentation or direct communication with the data provider. For the data set in the above example, it was originally obtained from online databases provided by the INFORMS organization,[1] and the detailed information is provided in the article by Ni et al. (2012), which can be accessed from the INOFRMS website as well.[2]

The documentation indicates that this dataset was derived from a retail store specializing in durable goods, such as televisions and other household appliances. In the original dataset (not shared here), individual household purchases

[1] https://pubsonline.informs.org/page/mksc/online-databases, accessed on June 3, 2025.

[2] https://pubsonline.informs.org/doi/abs/10.1287/mksc.1120.0726, accessed on June 3, 2025.

were recorded from December 1998 to November 2004, capturing details such as product category and purchase amount.

For this example, the original panel data (comprising multiple observations for each household) have been aggregated into a cross-sectional format, with one summary observation for each household. Each row represents a unique household, identified by `Household_ID`, and summarizes total spending and the number of purchases across various product categories (e.g., `E_cat2` indicating expenditure in category 2 and `N_cat4` indicating the number of purchases in category 4). Additionally, demographic information is provided for each household, including `Household_Income`. For illustration purposes, only a random sample from the original dataset is selected in this example.

With a solid understanding of the dataset's structure and content, we can now proceed with the analysis. The first step is to determine which variables should be included, based on the analysis's goals and the relevance of each variable.

Choose the Basis Variables

The basis variables are those selected for inclusion in the clustering analysis and are determined by the specific objective of the study. Choosing these variables is critical, as different sets of basis variables can produce very different segmentation results. To select the right variables, start with a clear view of the segmentation purpose and the key criteria that should matter when grouping customers. In this example, our goal is to segment customers based on their purchase behavior during store visits.

To achieve this, we need to include variables that reflect each customer's purchasing patterns across different product categories. Specifically, we will use all variables that begin with `E_categ`, as they capture total expenditures in each category and provide a meaningful behavioral profile for clustering.

Before proceeding with the analysis, please consider and answer the following question:

Q: Shall we also include the demographic variables in conducting the behavior-based segmentation analysis?

A: The answer is No, as we are trying to segment the customers based on their purchases (what they do), not based on their demographic information (who they are).

In the R code, we retrieve the names of the variables we wish to use for the clustering analysis.

```
basisvars <- names( dataset[, grep("E_", names(dataset)) ] )
#This command gets all the variable names that start with E_

basisvars <- basisvars[1:4] #use only the first few for demonstration
basisvars

## [1] "E_catg1" "E_catg2" "E_catg3" "E_catg4"
```

Creating a list that contains the names of variables to be used as the basis variable enhances both convenience and coding efficiency. This approach promotes good coding practices by allowing us to reference the list instead of repeatedly typing out each variable name. Examples will be demonstrated below.

Clean the Data

Before segmenting the customers in the data using their expenditure data, we need to clean the data.

Problems We May Encounter
When conducting clustering analysis, several potential issues may arise due to the nature of the recorded data:

- **Lack of variation in specific categories**: In some product categories, all customers may have spent the same amount. When a variable lacks variability, it provides no meaningful distinction across customers and, therefore, contributes little to the segmentation process.
- **Unequal scales across variables**: Some categories may have a much wider range of spending values than others. Standard clustering algorithms are sensitive to scale and may assign disproportionate weight to variables with larger ranges. This can distort the clustering results by overemphasizing categories that are not necessarily more informative about customer preferences.

To address these issues, data preprocessing steps such as variable selection and normalization or standardization are often necessary to ensure that all variables contribute appropriately to the analysis.

Let's examine these potential issues in our dataset.

```
diag(var(dataset[,basisvars]))

##   E_catg1   E_catg2   E_catg3   E_catg4
## 56335.701  9684.239     0.000  4400.142
```

From these results, we can see that category 3 has no variation in expenditures. For this reason, we should drop it from our basis variables:

```
basisvars = basisvars[-3]   # This removes the element in the third position of our list
basisvars

## [1] "E_catg1" "E_catg2" "E_catg4"
```

Note that the variances across product categories differ significantly. This variation is mainly due to differences in the scale or range of values for each category. For instance, the variance of `E_categ1` is more than ten times that of `E_categ4`. If we input the raw data directly into the clustering analysis algorithm, the segmentation results will be dominated by variables with larger variances by `E_categ1`, regardless of whether they are the most informative.

To prevent this imbalance, we must ensure that all variables are on a comparable scale before running the clustering analysis. This is achieved through **normalization**.

Normalization involves transforming each variable so that it has a mean of zero and a variance of one. This process standardizes the variables, allowing them to contribute equally to the clustering results, regardless of their original units or ranges.

By using these normalized variables, we can produce more balanced and meaningful customer segments based on behavioral similarity across all categories.

```
basisvars_n = paste0(basisvars, "_n")


# Assign normalized values to each of these new variables:
data_scaled = scale(dataset[,basisvars])
var(data_scaled)

##            E_catg1    E_catg2    E_catg4
## E_catg1 1.00000000 0.11641616 0.05746385
## E_catg2 0.11641616 1.00000000 0.04306944
## E_catg4 0.05746385 0.04306944 1.00000000

# Create a list of new variable names: "_n" for "normalized"
basisvars_n = paste0(basisvars, "_n")


# Assign normalized values to each of these new variables:
dataset[, basisvars_n] = scale(dataset[,basisvars])
```

After scaling the data, we expect the means to be 0 and the standard deviations to be 1 for each of the three columns.

```
# Check normalization
diag( var( dataset[, basisvars_n] ) )

## E_catg1_n E_catg2_n E_catg4_n
##         1         1         1

colMeans(dataset[,basisvars_n])

##     E_catg1_n     E_catg2_n     E_catg4_n
##  1.703853e-14  2.092998e-15 -1.220279e-16
```

All the variance values are 1, and all the mean values are close to 0, indicating that our normalization was successful.

Q: Why are these mean values not exactly 0?
A: These are numerical errors, but the values are essentially zero.

Performing Clustering Analysis

We are now ready to perform the clustering analysis. **Clustering** is a type of **unsupervised learning** in machine learning. Unlike **supervised learning**, such as regression analysis, where the goal is to predict a known outcome variable (Y) using a set of predictor variables (X), unsupervised learning does not rely on a predefined outcome variable; that is, there is no need for a Y variable. Instead, its objective is to uncover hidden patterns or groupings within the data.

In clustering analysis, we aim to identify individuals (e.g., customers) who are similar to one another based on a set of characteristics, represented by multiple variables, and group them into clusters. This is done using only the input variables (X), which describe each individual's features or behaviors. In this sense, clustering analysis can be thought of as a method that operates solely on multiple X variables, with no Y variable involved. If each customer can be represented by a single number, segmentation is straightforward: sort the customers by that value and define groups using cutoffs, such as top 30%, middle 40%, and bottom 30%. However, if each customer is represented by multiple characteristics, the task is much more complex, clustering analysis approach is required.

The three most commonly used clustering algorithms are:

- **K-Means Clustering**, which partitions the data into a predefined number of clusters by minimizing the variance within each cluster.
- **Hierarchical Clustering**, which builds a tree-like structure of nested clusters based on similarity, without requiring the number of clusters to be specified in advance.

- **DBSCAN** (Density-based spatial clustering of applications with noise) groups the data points that are closely packed and similar into the same cluster while keeping the low-density regions as noise or outliers.

Each method has its strengths and applications, and both will be introduced in the following sections.

K-Means Method for Clustering Analysis

K-Means is a popular solution method for conducting clustering analysis.

Understand the K-Means Method

The K-means clustering algorithm is widely used for partitioning data into K distinct groups, where K is a predefined number. The algorithm follows these general steps:

1. **Initialization**: Each data point is randomly assigned to one of the K clusters.
2. **Centroid Calculation**: For each cluster, calculate the centroid, which is the mean position of all members in that cluster (in multidimensional space).
3. **Reassignment**: Reassign each data point to the cluster whose centroid it is closest to. This updates the group memberships.
4. **Update Centroids**: Recalculate the centroids for the new clusters.
5. **Repeat**: Steps 3 and 4 are repeated until the group assignments no longer change, that is, until convergence.

The final result assigns each data point to the cluster with the closest centroid, forming K groups that minimize within-cluster variance.

It's important to note that K-means provides a **heuristic solution** to a complex problem. While it often yields good results quickly, it does not guarantee finding the best solution. Due to the random initialization of clusters, different runs of the algorithm can produce varying results, especially when the data contains overlapping or ambiguous group boundaries.

Although achieving the accurate optimal clustering solution may require more complex and time-consuming computations, K-means offers a computationally efficient and scalable solution that is particularly useful when working with large datasets. This balance between accuracy and efficiency is one reason why K-means remains a popular choice for clustering in the practice of customer segmentation.

Implementing K-Means in R

The ***way to conduct*** The K-means() command conducts the K-means analysis clustering analysis in R. To apply the K-means method, you first need to define the number of groups/segments you want, which is K as referred to in the name.

```
# Run k-means command with three groups
km <- kmeans( dataset[,basisvars_n], 3)
# Add the k-means classifications to our data.table
dataset[,'seg']= km$cluster
```

We can examine the size of each segment and the mean values of each segment for each of the variables used in segmentation.

```
cbind(km$size,km$centers)

##           E_catg1_n  E_catg2_n   E_catg4_n
## 1   451  0.4161566  5.6208059  0.17070222
## 2 18337 -0.1991343 -0.1337508 -0.02298751
## 3  1148  3.0172814 -0.0717733  0.30011789
```

As discussed earlier, the results from K-means are not always stable; the results will vary each time it is run. For example, the following two results are from running the same K-means two times.

```
       E_catg1_n  E_catg2_n   E_catg4_n
1 18329 -0.1997590 -0.13387353 -0.02329431
2   454  0.4114685  5.58820585  0.49080005
3  1153  3.0135099 -0.07222676  0.17704967
```

First K-means Result

```
       E_catg1_n  E_catg2_n   E_catg4_n
1 19300 -0.0183415 -0.1328747 -0.06863341
2   176  0.4997529  0.0716446  7.38839317
3   460  0.5783356  5.5475472  0.05275554
```

Second K-means Result

To ensure the computer program will give out repeatable results, you can set the random variable seed before running the `kmeans()` analysis. The results will be the same any time you run the code below, unless you change the seed number.

```
set.seed(123)
print("k-means resutls with a different seed")

## [1] "k-means resutls with a different seed"

km <- kmeans( dataset[,basisvars_n], 3)
cbind(km$size,km$centers)

##           E_catg1_n   E_catg2_n   E_catg4_n
## 1 18329 -0.1997590 -0.13387353 -0.02329431
## 2   454  0.4114685  5.58820585  0.49080005
## 3  1153  3.0135099 -0.07222676  0.17704967

set.seed(321)
print("k-means resutls with one seed")

## [1] "k-means resutls with one seed"

km <- kmeans( dataset[,basisvars_n], 3)
cbind(km$size,km$centers)

##           E_catg1_n  E_catg2_n   E_catg4_n
## 1  1148  3.0172814 -0.0717733  0.30011789
## 2 18337 -0.1991343 -0.1337508 -0.02298751
## 3   451  0.4161566  5.6208059  0.17070222
```

How to Choose K

The clustering analysis above was performed using $K = 3$ segments. But how do we know if 3 is the correct number of clusters? Selecting the optimal number of segments is a crucial step in the clustering analysis process.

A common and practical approach is to try several different values of K and compare the results using a quantitative metric. One widely used metric is the Within-Cluster Sum of Squared Errors (within-SSE). This metric is conceptually similar to the total within-group variance. It measures the sum of the squared distances between each data point and the centroid of its assigned cluster, aggregated across all clusters.

As K increases, the number of points in each cluster generally decreases, leading to more homogeneous groupings. This naturally reduces the within-SSE. However, the improvement diminishes with each additional cluster. Eventually, the gains from increasing K become smaller, to the point that it is not worthwhile to reduce the within-SSE anymore, given the cost associated with a large number of segments K, in practice.

To identify the optimal K, we look for the point at which the reduction in within-SSE levels off, which is a method known as the **elbow method**. This is best visualized by plotting the within-SSE against various values of K. The "elbow point" on the plot refers to the point where the curve begins to flatten. This point suggests a good balance between model complexity and explanatory power.

The code below generates the elbow plot to help guide the selection of the optimal number of clusters.

```
maxnum = 7

# Create a data.frame to stall the SSE for each number of clusters
winss = data.frame('Clusters'=c(1:maxnum),'SSE'=rep(0,maxnum))
for (i in 1:maxnum)
  winss[i,2] = kmeans(dataset[,basisvars_n], i)$tot.withinss

# Make the plot and find the "elbow"
ggplot(winss, aes(x=Clusters, y=SSE, group)) +
  geom_line() +
  geom_point()
```

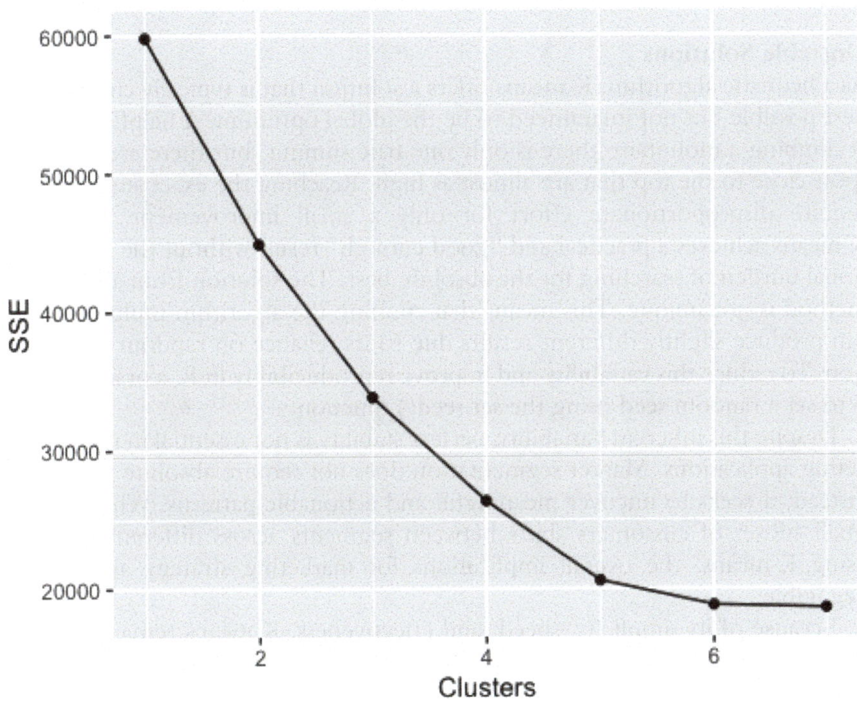

From this plot, the reduction of the metric slows down after $K = 5$. As a result, we can choose 5 to be the desirable K value. This method is also called the elbow method.

Limitations of the K-Means Method

This method is simple to implement and quite intuitive. However, it does have its limitations:

The Number of Segments Needs to Be Predetermined

The number of segments (the *k*-value) must be determined before conducting the analysis. Fortunately, in marketing segmentation, having too many segments is impractical to manage, as marketers cannot effectively manage an excessive number of distinct customer groups. In practice, a typical number of customer segments ranges from 3 to 7 segments. To achieve the best result, it is recommended to run the K-means() analysis multiple times, each time with a fixed number of segments, and then compare the results to select the one with the most suitable outcome.

Unstable Solutions

As a heuristic algorithm, K-means offers a solution that is typically close to the best possible but not guaranteed to be the global optimum. A helpful analogy is climbing a mountain: there is only one true summit, but there are multiple spots close to the top that are almost as high. Reaching the exact summit may require disproportionate effort for only a small improvement. Similarly, K-means achieves a practical and "good enough" result without the computational burden of searching for the absolute best. The solution from a K-means method is not unique. This means that running the algorithm multiple times can produce slightly different results due to its reliance on random initialization. To reduce this variability and improve reproducibility in R, a practical step is to set a random seed using the set.seed() function.

Despite this inherent variability, perfect stability is not essential in most marketing applications. Market segmentation does not require absolute precision; instead, it seeks to uncover meaningful and actionable patterns. When only a small subset of customers shifts between segments across different analyses using K-means, the overall implications for marketing strategy are usually negligible.

Because of its simplicity, speed, and effectiveness, K-means remains one of the most widely used methods for behavioral segmentation, particularly when working with large datasets.

However, in situations where greater precision or consistency is required, such as in smaller datasets or strategic contexts where customer grouping needs to be highly reliable, hierarchical clustering may be a better alternative. Unlike K-means, hierarchical clustering does not rely on heuristic approaches; instead, it provides a solution based on a nested grouping structure.

Hierarchical Clustering

Similar to the K-means method, hierarchical clustering (HC) is also an unsupervised learning model. It identifies similar items and categorizes them into the same group, while keeping distinct items in separate groups.

Here is a description of this approach:

In hierarchical clustering (HC), the core idea is to group individuals who are similar and are close in distance into the same cluster. To determine similarity, we must first define a distance metric between each pair of customers. The most commonly used metric is the Euclidean distance.

Let's denote the set of variables that describe the customer i as x_{i1}, x_{i2}, ..., x_{i15}. For example:

- x_{i1}: Number of products purchased by customer i in category 1
- x_{i2}: Number of products purchased in category 2
- ...
- x_{i15}: Number of products purchased in category 15

Similarly, for the customer j:

- x_{j1}, x_{j2}, ..., x_{j15}: Number of products purchased in each corresponding category

The **Euclidean distance** between customers i and j is computed as:

$$d_{ij} = \sqrt{\left(x_{i1} - x_{j1}\right)^2 + \left(x_{i2} - x_{j2}\right)^2 + \ldots + \left(x_{i15} - x_{j15}\right)^2}$$

This formula calculates the straight-line distance between the two customers in a 15-dimensional space, where each dimension represents a purchase activity in a specific product category. Customers with more similar purchase behaviors will have a smaller Euclidean distance between them, making them more likely to be grouped into the same cluster in the hierarchical clustering process.

In the example dataset, there are $N = 19{,}936$ customers. To perform hierarchical clustering, we need to compute the pairwise distances between all possible pairs of customers. This results in:

$$\frac{N \times (N-1)}{2} = \frac{19{,}936 \times 19{,}935}{2} \approx 198.7\,\text{million}$$

This imposes a significant computational burden, especially for large datasets. The hierarchical clustering process proceeds as follows:

1. **Initialization**: Begin with each customer in their own segment, resulting in $N = 19{,}936$ segments.
2. **First Merge**: Identify the pair of customers (or clusters) with the shortest distance and merge them, reducing the number of segments to $N - 1$.

3. **Iterative Merging**: At each subsequent step, merge the two clusters with the smallest within-group sum of squares (ESS) or minimum distance, depending on the distance metric.
4. **Repeat**: Continue merging until all customers are combined into a single cluster.

Due to the scale of the data, visualizing the complete clustering process is impractical. Instead, we illustrate the method using a subset of customers. The R code below demonstrates how to compute hierarchical clustering and visualize the results using data from the first few customers.

```
# calculate the distance among households with an ID less than a number
# to limit the number of customers we are dealing with
sum(dataset$Household_ID<100090000) #number of customers selected

## [1] 44
d=dist(dataset[dataset$Household_ID<100090000,basisvars_n],method="euclidian"
)

# Get the dendrogram to visualize the results from running this hierarchical
clustering analysis
fit = hclust(d,method="ward.D")
plot(fit)
```

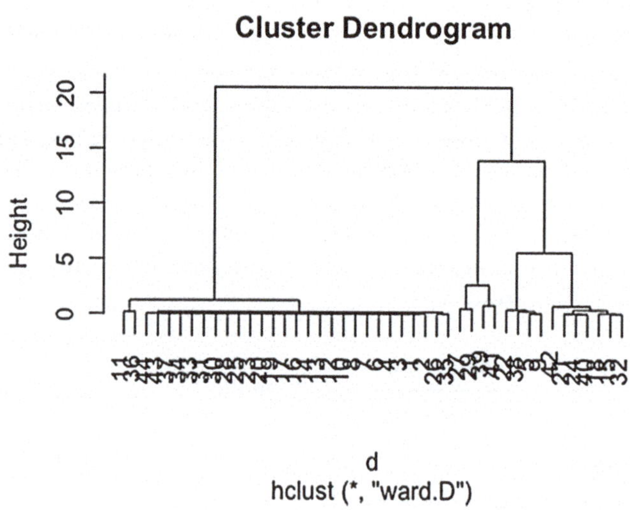

In the hclust() function, you can choose different methods to cluster data based on distance.

Ward.D is one of the most popular methods adopted in hierarchical clustering analysis, which identifies clusters with the minimum variance within them.

To determine the number of segments, we can examine the height either from the plot or using the following function

```
fit$height
##  [1] 0.000000e+00 0.000000e+00 0.000000e+00 0.000000e+00 0.000000e+00
##  [6] 0.000000e+00 0.000000e+00 0.000000e+00 0.000000e+00 0.000000e+00
## [11] 0.000000e+00 0.000000e+00 0.000000e+00 0.000000e+00 0.000000e+00
## [16] 0.000000e+00 0.000000e+00 0.000000e+00 0.000000e+00 0.000000e+00
## [21] 0.000000e+00 0.000000e+00 0.000000e+00 0.000000e+00 0.000000e+00
## [26] 1.263949e-04 5.940558e-03 5.029111e-02 6.323956e-02 8.409471e-02
## [31] 1.431632e-01 1.555359e-01 2.456133e-01 3.745360e-01 3.864031e-01
## [36] 4.877626e-01 7.019950e-01 7.823509e-01 1.209132e+00 2.647484e+00
## [41] 5.593872e+00 1.386711e+01 2.055416e+01
```

Find a really high jump in the plot, and then draw a horizontal line. The number of vertical lines crossing this line defines the number of segments. For example, if you think that from 2.65 to 5.59 is a huge jump, which should not have happened. We draw a horizontal line between these two numbers in the dendrogram; the line interacts with three vertical lines. It means we chose the number of segments to be 3. To get the membership of each customer in 3 segments, we can use

```
fitmem = cutree(fit,k=3)
print(fitmem)

##  1  2  3  4  5  6  7  8  9 10 11 12 13 14 15 16 17 18 19 20 21 22 23 24 25
 26
##  1  1  1  1  2  1  1  1  2  1  1  1  1  1  2  1  1  2  1  1  2  2  1  2  1
  1
## 27 28 29 30 31 32 33 34 35 36 37 38 39 40 41 42 43 44
##  3  1  3  1  1  2  1  1  1  1  1  2  3  2  3  2  1  1

addindx=cbind(seq(1:44),fitmem)
print(t(addindx))

##        1 2 3 4 5 6 7 8 9 10 11 12 13 14 15 16 17 18 19 20 21 22 23 24 25 2
6 27
##        1 2 3 4 5 6 7 8 9 10 11 12 13 14 15 16 17 18 19 20 21 22 23 24 25 2
6 27
## fitmem 1 1 1 1 2 1 1 1 2  1  1  1  1  1  2  1  1  2  1  1  2  2  1  2  1
1  3
##        28 29 30 31 32 33 34 35 36 37 38 39 40 41 42 43 44
##        28 29 30 31 32 33 34 35 36 37 38 39 40 41 42 43 44
## fitmem  1  3  1  1  2  1  1  1  1  1  2  3  2  3  2  1  1
```

It prints out the segment each member is assigned to based on this segmentation analysis.

We won't be able to illustrate the HC using the entire data set, as it would take too long to complete running the HC model. This is one of the most critical limitations of the HC approach: it is computationally very costly. Its obvious benefits include:

- It does not require a predefined number of segments.
- It has a unique solution.

However, while market segmentation using customer-level data does not require an exact solution, it typically involves large datasets. This makes hierarchical clustering (HC) impractical for datasets of meaningful size due to its high computational complexity.

DBSCAN

Another common approach for market segmentation is DBSCAN, which stands for Density-Based Spatial Clustering of Applications with Noise. This approach identifies data patterns that are similar among the high-dimensional data and clusters them together. Unlike hierarchical clustering, it does not require as much computational power. Unlike the K-means method, it does not require predefining the number of segments.

Intuitively, it involves the following steps:

- It starts with a random data point, representing one customer, with high-dimensional data.
- It finds several nearest neighbors, represented by other customers that are closest to the focal customer. The difference between any two customers is evaluated by the distance between these two high-dimensional data points.
- For each neighbor, it continues to find a certain number of other customers that are similar to the neighbor. All of these "similar" customers are grouped into one segment. When not enough neighbors can be added, it will start the second segment with another data point that has not been included in the current cluster.

In implementing this algorithm, two values need to be predefined:

1. eps (ϵ): specifies the radius around a point used to search for neighbors. The value of this parameter determines the number of clusters to be formed. If eps is set too high, many customers who are not very similar may be grouped into the same segment. If it is set too low, the algorithm may produce too many small segments.
2. minPnts: specifies the minimum number of points required for a region within the eps-radius to be considered dense enough to form a cluster. If a data point finds fewer than minPnts neighbors within eps-range, it may be a border point (if close to another data point included in a cluster) or noise (if isolated).

To calculate the distance between any two points (representing two customers), Euclidean distance is often the default, although other distance metrics can be used.

For illustrative purposes, we select the first 10 categories of expenditures as variables for the segmentation analysis in this exercise.

```
basisvars <- names( dataset[, grep("E_", names(dataset)) ] )
#This command gets all the variable names that start with E_

basisvars <- basisvars[1:10] #use only the first few for demonstration
basisvars

##  [1] "E_catg1"  "E_catg2"  "E_catg3"  "E_catg4"  "E_catg5"  "E_catg6"
##  [7] "E_catg7"  "E_catg8"  "E_catg9"  "E_catg10"
```

Similar to what we illustrated before, we remove the third category, as it has no data. Then, the data is scaled so that each column has a variance of 1 and a mean of 0, before we conduct the analysis.

```
basisvars = basisvars[-3]   # This removes the element in the third position
of our list
basisvars_n = paste0(basisvars, "_n")

# Assign normalized values to each of these new variables:
dataset[, basisvars_n] = scale(dataset[,basisvars])
diag( var( dataset[, basisvars_n] ) )

##  E_catg1_n  E_catg2_n  E_catg4_n  E_catg5_n  E_catg6_n  E_catg7_n  E_catg8
_n
##          1          1          1          1          1          1
 1
##  E_catg9_n E_catg10_n
##          1          1

colMeans(dataset[,basisvars_n])

##      E_catg1_n     E_catg2_n     E_catg4_n     E_catg5_n     E_catg6_n
## -5.128250e-18 -3.398965e-18  9.781139e-18 -1.636745e-17 -3.902432e-18
##      E_catg7_n     E_catg8_n     E_catg9_n    E_catg10_n
##  1.191256e-17 -7.677474e-18  2.623682e-17 -7.792333e-18
```

Load the library for DBSCAN.

```
#install.packages(c("dbscan","dplyr"))  # if needed
library(dbscan)
library(dplyr)
```

For illustration, we use the first 5000 data points in the data.

```
knnid = 1:5000  # number of customers to sample for clustering
data_dbscan = dataset[knnid,basisvars_n]
data_dbscan[is.na(data_dbscan)] <- 0  # DBSCAN can't handle NAs]
```

Specify `minPnts`

Specify the value of `minPnts`, which indicates the minimum number of neighbors required for a data point to be considered in a dense enough region. It is common to define this value as twice the dimension of the data.

```
# Rule of thumb for minPts: ~ 2 * dim
minPts <- 2 * ncol(data_dbscan)
minPts

## [1] 18
```

Determine the `eps` *Value*

Based on the results from the chosen `minPnts`, we calculate the distance between each point and its `minPnts` nearest neighbor using the function `kNNdist()`.

The value of `eps` is the 95th percentile of all these nearest distances for all the data points included in the analysis.

```
# Compute the distance to the k-th nearest neighbor for each point
kNN <- kNNdist(data_dbscan, k = minPts)

# Take the 95th percentile:
eps_guess <- as.numeric(quantile(kNN, 0.95))
eps_guess

## [1] 2.598579
```

Run DBSCAN

With both parameters specified, we can cluster the data using the `dbscan()` function. The number of customers in each segment is calculated using function `table().`. Those belonging to segment 0 are outliers classified by the algorithm.

```
# Fit DBSCAN
db <- dbscan(data_dbscan, eps = eps_guess, minPts = minPts)

# Attach cluster labels to customers (0 is noise)
data_dbscan$cluster <- factor(db$cluster)

# Quick diagnostics
table(data_dbscan$cluster)  # counts per cluster (including noise)

##
##    0    1    2
##  140 4827   33
```

Show the Average for Each Segment

Calculate the average value of each of the expenditure variables among those within the same cluster.

```
cluster_mean <- data_dbscan %>%
  group_by(cluster) %>%
  summarise(
    n = n(),
    across(all_of(basisvars_n), ~mean(.x, na.rm = TRUE))
  )
cluster_mean

## # A tibble: 3 × 11
##   cluster     n E_catg1_n E_catg2_n E_catg4_n E_catg5_n E_catg6_n E_catg7_
n
##   <fct>   <int>     <dbl>     <dbl>     <dbl>     <dbl>     <dbl>     <dbl
>
## 1 0         140    2.33      1.92      1.12      1.46      1.93      1.88

## 2 1        4827   -0.0633   -0.0471   -0.0699   -0.0355   -0.0609   -0.077
0
## 3 2          33    0.0453   -0.118     5.91      0.0528   -0.0849    0.122

## # ℹ 3 more variables: E_catg8_n <dbl>, E_catg9_n <dbl>, E_catg10_n <dbl>
```

Comparisons Between K-Means and DBSCAN

Although both methods are heuristic approaches, they are different in many ways.

First, the K-means method requires predefining only one number, which is the number of segments. In contrast, DBSCAN requires predefining two values: the minimum number of neighbors and the radius. To find suitable values, both methods may require a trial-and-error approach.

Second, one of the outputs from the DBSCAN approach is segment 0, which consists of the outliers. It may require further investigation before deciding whether it requires further segmentation among those belonging to segment 0.

Summary

Customer segmentation is one of the most fundamental tasks in marketing. Traditionally, segmentation has been based on:

- **Demographics**: information about who the customers are (e.g., age, gender, income)
- **Geographics**: information about where they are located
- **Psychographics**: information about their lifestyles, values, and interests, often gathered through surveys such as VALS

With the rise of digital technologies, companies now have access to detailed tracking data on individual customer interactions with websites, digital ads, and other online channels. These behavioral data provide a more direct and dynamic view of customer preferences for products and services. Leveraging such data enables more precise, behavior-based segmentation, which in turn facilitates more effective targeting strategies.

To conduct behavioral segmentation, two commonly used clustering techniques are:

- **K-Means Clustering**: Efficient and scalable for large datasets. However, the solution may vary across runs, and the number of segments must be specified in advance.
- **Hierarchical Clustering**: Produces a consistent result without requiring a predefined number of clusters. However, it is computationally intensive and less suitable for large datasets.
- **DBSCAN**: Efficient and no need to pre-specify the number of segments. Instead, it requires defining two metrics that influence the segmentation results: eps (the neighborhood radius) and minPnts (the minimum number of neighbors). A data point with at least minPnts neighbors within the eps-range is considered to be in a dense region and is part of a cluster.

In practice, both DBSCAN and K-means are commonly used in marketing for segmentation purposes due to their scalability and efficiency.

Homework

1. Load the durable goods data file DurData_custpurchase.csv.
2. Use all the variables starting with "N_catg" as the basis variables.
3. Perform customer segmentation in two stages:
    (a) First, apply the K-means clustering approach, making sure to normalize the data first.
    (b) Use the elbow method to determine the number of clusters (choose among 3, 4, or 5 segments).
    (c) Next, conduct the second-round segmentation, using DBSCAN for comparison. Comment on the differences in the results.

Reference

Ni, Jian, Scott A. Neslin, Baohong Sun, (2012), "Database Submission—The ISMS Durable Goods Data Sets," *Marketing Science*, 31(6):1008–1013.

CHAPTER 6

Targeting and Propensity Score Model (Part I)

Learning Objectives

1. Explain the various types of targeting strategies used in marketing analytics.
2. Apply the propensity score approach to design an effective targeting strategy.
3. Implement a binary logit model and interpret its results.
4. Compare the differences between linear regression and logistic regression models.

Introduction

Targeting refers to the strategic process of identifying specific segments of customers who are most likely to be interested in a product or service. This practice enables businesses to tailor their marketing efforts to meet the unique preferences and behaviors of different customers, thereby improving the effectiveness of marketing campaigns.

Targeting is crucial because it allows companies to allocate resources more effectively and enhance customer satisfaction by delivering relevant and personalized experiences. In modern marketing, targeting is implemented based on customer-level data, such as browsing history, purchase behavior, and demographic information. This data plays a critical role in optimizing marketing strategies. By leveraging data-driven insights, businesses can create highly personalized marketing messages, improve customer engagement, and increase purchase probabilities. Moreover, precise targeting reduces wasted advertising spend and ensures that marketing efforts reach the right audience at the right time, ultimately driving business growth and competitive advantage.

Supplementary Information The online version contains supplementary material available at https://doi.org/10.1007/978-3-032-11130-2_6.

© The Author(s), under exclusive license to Springer Nature Switzerland AG 2026
X. Dong, *Marketing Analytics and Data Science*,
https://doi.org/10.1007/978-3-032-11130-2_6

The concept and processes of leveraging customer-level data to build targeting models are explored in the following two chapters. This chapter focuses on the model-building process, and the next chapter completes the data-driven targeting approach by addressing common challenges in data pattern and model comparisons.

Different Types of Targeting

There are various ways to classify targeting strategies. In this chapter, we introduce a classification method based on two dimensions of consideration. This classification framework helps analysts make informed modeling choices when implementing data-driven targeting strategies.

Types of Targeting Strategies

In this classification framework, we consider two dimensions:

Dimension A: by level of granularity

- A1: Targeting a selection of segments, treating each group of customers differently. In traditional marketing practices, this may be the only feasible targeting strategy due to the lack of individual-level data. It remains popular today due to its cost efficiency.
- A2: Targeting a selection of individual customers, treating each customer differently. This is less common and requires more data tracking and analysis.

Dimension B: by goal

- B1: Targeting customers who are likely to purchase.
- B2: Targeting customers who are likely to respond to a specific marketing action.

Any targeting strategy can be regarded as a combination of the above two dimensions, resulting in four possible combinations (Fig. 6.1).

The detailed considerations for each dimension are explored below.

Dimension A: Segment- Versus Individual-Level Targeting

In implementing targeting strategies, business tries to treat different customers differently. The most advanced level of targeting involves identifying the differences among individual customers and applying a tailored approach to each one. However, this level of personalization is not always feasible. Instead, we can segment customers into groups (segments) and apply differentiated strategies to each group while treating all those in the same group similarly.

	B1: Targeting to Buy	B2: Targeting to Promote
A1: Segment level targeting	Targeting to the groups of customers who are likely to buy. *For example, when Tesla Model S first came out, it targets to high income individual who are techies.*	Targeting to the groups of customers who are likely to respond. *For example, United Airlines send a mailer to only their Gold members, based on the loyalty status.*
A2: Individual level targeting	Targeting to the individual customers who are likely to buy. *For example, Amazon recommended products after you make a purchase, indicating "people who bought that product also purchased the following products."*	Targeting to individual customers who are likely to respond to marketing offerings. *For example, Walmart would like to send their coupons to only those who would not have bought without the coupons.*

Fig. 6.1 Four types of targeting strategies

Segment-level targeting is more common, as it is less costly and more practical in many business environments. For example, automakers often offer multiple product lines under different brands, targeting customers based on income levels or life stages. In this approach, Toyota markets Lexus sedans and SUVs to high-income customers, while the Sienna minivan is designed for families.

Individual-level targeting is gaining popularity as the ability to track customer-level data improves and analytics technologies advance. For instance, Amazon personalizes its webpages by recommending products based on a user's past browsing and purchase history, enhancing the customer experience through data-driven personalization.

Comparisons

Segment-level targeting:

- Pro: Less costly and easier to implement.
- Con: Not consistently effective as it ignores the differences among the customers within the same segment.

Individual level targeting:

- Pro: More costly and sometimes impossible to implement.
- Con: Requires more data and advanced analytical technologies in building models. It may also run into hurdles due to user-level data privacy regulations (Fig. 6.2).

To implement individual-level targeting effectively, two key requirements must be met:

	Implementation Cost	Potential Benefits
Segment level targeting	Lower Easier to implement	Lower Treating everyone in the same segment the same.
Individual level targeting	Higher Not always feasible to reach each customer individually	Higher Treating everyone differently with different marketing offerings.

Fig. 6.2 Comparison of segment-level versus individual-level of targeting

1. Detailed Customer Information: Companies need comprehensive data on each customer, including demographics, purchase history, and online behavior. This information enables the development of personalized marketing strategies tailored to individual needs.
2. Feasible Communication Channels: Companies must have efficient methods to reach each customer individually, such as personalized emails, targeted digital ads, or customized website experiences. Advances in data analytics and automation have made it increasingly practical to deliver such personalized interactions at scale.

In the meantime, companies are also working diligently to ensure compliance with privacy regulations, thereby protecting their customers' information and earning their trust. These new regulations, most notably the GDPR by the European Union and the CCPA by the state of California, create additional challenges for companies in implementing individual-level targeting.

Dimension B: Targeting to Buy or Promote

Targeting to Buy

One common marketing objective is to identify customers who are most likely to make a purchase. To do this, we can build a propensity score model that predicts the probability of purchase as a function of a set of observed variables, including variables describing each customer in terms of who they are and what they did, in addition to the variables recording marketing actions taken by the company at different time periods and targeted towards different customers.

Let:

- Y: whether to buy ($Y = 1$) or not ($Y = 0$)
- X: a set of factors that influences the purchase decision, which includes the following:
  - Customer interactions with the company, what they did (e.g., website visits, prior purchases)

- Demographic information, who they are (e.g., age, income, location)
- Marketing treatments received (e.g., emails, coupons)

The model specification follows a regression framework:

$$\text{Prob}(Y=1) \sim \beta_0 + \beta_1 X_1 + \ldots + \beta_k X_k + \epsilon \tag{6.1}$$

The targeting-to-buy strategy refers to identifying customers with the highest probability of purchase, which is predicted using the above model.

Targeting to Promote

In promotion targeting, the goal is to identify customers who are most responsive to a specific marketing action, such as a coupon. A straightforward approach is to estimate a model where one of the predictors (e.g., X_1) represents the marketing action of interest. Customers with higher estimated values of $\widehat{\beta_1}$ would then be prioritized for the promotion.

However, a limitation of this approach is that $\widehat{\beta_1}$ is constant across all individuals. This means we cannot differentiate which customers are more or less responsive to the particular promotion represented with the variable X_1.

To overcome this, we can evaluate heterogeneous effects using one of two methods in our modeling approach:

1. **Interaction Terms**: Introduce interaction variables between demographics and the marketing treatment. For example, if X_1 is the coupon and Z is age, including $X_1 \times Z$ in the model. The estimated effect of the coupon becomes $\widehat{\beta_1} \times \text{age}$, which allows the promotional impact to vary by the customer's age.
2. **Hierarchical Bayesian Modeling**: This approach allows the coefficients on all the model variables, including the promotion variable (e.g., coupon) to vary across individuals without being constrained by observed demographic variables. It provides a flexible framework for modeling individual-level responsiveness. However, this method is beyond the scope of this course. We will introduce the basic concepts of Bayesian modeling and its applications in marketing analytics in Chaps. 9 and 10. For those interested in more about Bayesian approach, including Hierarchical Bayesian modeling, we recommend the book *Bayesian Statistics and Marketing* by Rossi, Allenby, and McCulloch.

In summary:

- **Targeting to buy** focuses on identifying high-probability purchasers using a **propensity score** model.
- **Targeting to promote** focuses on identifying highly responsive customers using interaction terms or more advanced techniques like hierarchical Bayesian models.

Combine the Two Dimensions

Among the four targeting strategies discussed above, the two that focus on **individual-level targeting** have gained popularity in practice. This is mainly due to the increased availability of customer-level data and advancements in analytical tools, which make these approaches more efficient and effective compared to segment-level strategies.

Of the two individual-level approaches—targeting to purchase and targeting to promote—this chapter focuses on an example of targeting to buy. However, the same modeling framework can also be applied to targeting to promote, through the application of one of two methods to capture individual responsiveness: incorporating interaction terms or employing a Hierarchical Bayesian approach.

Targeting with Propensity Scores

In this session, we explore how to build a model that identifies the right individuals to target based on their predicted probability of making a purchase or taking some actions that are desired by the marketer, such as service subscription, new member signups or ads clicks. To achieve this, we need to develop a propensity score model.

The propensity score is an important concept in Marketing Analytics, and is commonly used to predict the likelihood of a customer taking a specific action. These actions are typically those that directly impact a company's success and reflect the desired outcomes of marketing decisions, for example:

- propensity score for making a purchase
- propensity score for opening an email
- propensity score for clicking on an advertisement
- propensity score for writing a product review
- propensity score for churn (stopping subscription to a service)
- ……

Various machine learning techniques can be applied to develop a propensity score model, including linear regression, support vector machines, random forests, and many more. However, in this chapter, we focus on one of the most popular approach, logistic regression. Logistic Regression model can effectively model and predicts the probabilities and, more importantly, provide interpretable insights. Many of the concepts discussed here also apply to the other modeling approaches mentioned above.

Based on the targeting classifications discussed earlier, this task falls under the combination of "individual level" and "targeting to buy".

Since this strategy relies on predicting each customer's behavior, it is also known as behavior-based individual-level targeting. Like behavioral-based segmentation, this approach leverages customer data, particularly behavioral records, to make informed decisions. However, unlike traditional segmentation and targeting methods, which primarily rely on demographic information, behavior-based approaches focus on observed actions and engagement patterns to understand customers and provide suggestions for effective targeting strategies.

In logistic regression, the *Y* variable takes binary values (0 or 1) and is also known as the Binary Logit model. The selection of the *X* variables is very similar to that in building a linear regression model.

To illustrate this approach, we collaborate with the Development Office for an unnamed university, which maintains a database of their alumni. Each year, during the Grand Reunion, alumni who graduated in multiples of five years (e.g., 5th, 10th, 15th reunion) are invited back to campus for a series of parties and events. The alumni office aims to leverage this opportunity to encourage donations to the university from attendees.

To maximize the effectiveness of their outreach while minimizing costs and avoiding unnecessary mailers, the development office seeks to identify alumni who are most likely to donate. To accomplish this, we build a propensity score model that predicts each individual's probability of donating this year.

To build a propensity score model predicting each alum's probability of donation, we need the following information in the data:

- The dependent variable *Y*, indicating whether each alum has donated to the school in the past.
- The *X* variables that may have influenced the outcomes of the *Y* values.
  - Who they are: demographics of each individual
  - What they did: the activities they have participated in, both before and after graduateion, and their actions taken in response to the past marketing efforts from the development office.

To build a propensity score model, the process involves three steps:

- Step 1: Develop a propensity score model using the past donation information among the alumni.
- Step 2: Based on the model training results, predict each alumnus or alumna's probability of donating by calculating their propensity score for donation.
- Step 3: Send the promotion materials to those who are predicted to have the highest propensity to donate.

The details of each step are explored in the following sections.

Step 1: Build a Logistic Regression Model

The setup of a logistic regression model is very similar to that of a linear regression model, which was discussed in Chap. 4.

The R code needed to implement the analysis is provided here. The equivalent Python code is available online as supplementary material.

First, Load the Data

```
rm(list=ls())        # Remove everything in memory to start fresh.
library(MASS)        # Load the MASS package

reuniondata <- read.csv('Reuniondata_inclass.csv')
head(reuniondata)

##   CaseNum RYCohort donatesum SpouseAlum SportsAlum UGAlumAwards OtherUGAct
## 1       1     1969         0          0          1            0          0
## 2       2     1964         0          0          0            0          0
## 3       3     1964         0          0          0            0          0
## 4       4     1964         1          0          0            0          0
## 5       5     1959         0          0          0            0          0
## 6       6     1959         0          0          0            0          0
##   EverAssigned BoardMember ChildAlum ParentAlum SiblingAlum GradDegree
## 1            0           0         0          0           0          0
## 2            0           0         0          0           0          0
## 3            0           0         0          0           0          0
## 4            1           0         0          0           0          0
## 5            0           0         0          0           0          0
## 6            0           0         0          0           0          0
##   TotalReunions OnePlusEvents Years.Lapsed
## 1             0             0           48
## 2             0             0           32
## 3             0             0           48
## 4             2             1            4
## 5             0             0           20
## 6             0             0           48
```

Second, Understand the Data

The data recorded some information about each alum and the values they have donated so far.

The information about each alum includes:

- RYcohort: the year of graduation.
- donatesum: total value of donations.
- SpouseAlum: whether spouse is also an alum.
- SportsAlum: whether the student was in a sports team while at SCU.
- UGalumAwards: whether the student won some awards as an undergraduate student at SCU.
- OtherUGAct: whether the student participated in other activities as an undergraduate student.

- EverAssigned: whether the alumnus was ever assigned an officer from the development office. Only those with significant wealth are likely to be assigned an officer.
- BoardMember: whether the alumnus has ever been on the board.
- ChildAlum: whether the alumnus has a child who attended SCU.
- ParentAlum: whether the alumnus has a parent who attended SCU.
- SiblingAlum: whether the alumnus has a sibling who attended SCU.
- GradDegree: whether the alumnus has a graduate degree from SCU.
- TotalReunions: Total number of reunions participated in by the alumnus.
- OnePlusEvents: whether the alumnus has attended one of those OnePlus events, one of the fundraising events.
- Years.Lapsed: number of years since last donation.

Third, Define the Y *Variable*

Given that we are interested in predicting the donation probability, the dependent variable Y is defined as whether the person donated or not.

To achieve this, we use the variable "donatesum" to represent the total donation; if it is 0, it means the person has not donated; otherwise, they have donated.

Create one more column in the data, named "Choice". As a result, the number of columns in the data increased from 16 to 17.

```
reuniondata[,"Choice"]=as.numeric(reuniondata[,"donatesum"]>0)
dim(reuniondata)

## [1] 7120   17
```

Fourth, Define the X *Variables and Estimate the Model*

As a start, we include all the other variables in the data frame as X variables in the model.

To estimate the logistic regression model, we need to use the R function glm().

In comparison, recall that in estimating a regression model, we use the R command lm().

Some explanations of the R code below:

- glm() can be used to estimate a variety of generalized linear models. To specify that it is a binary logit model, we use the family="binomial" argument.
- In this example, the formula of the model is written as Choice~ ., which indicates that we use the column data_use[,"Choice"] as the Y variable, and the rest of the columns in the data.frame as the X variables.

Similar to a regression model, we use the function summary() to display the model estimation results.

```
# First, create a data.frame with only the variable we would use,
# by dropping the first three columns, including
# 1. case number
# 2. Year of the cohort
# 3. total value of donations
data_use <- reuniondata[,-c(1:3)]

# Then, estimate the binary logit (BL) model
bl_result = glm(Choice~., data=data_use, family="binomial")
summary(bl_result)

##
## Call:
## glm(formula = Choice ~ ., family = "binomial", data = data_use)
##
## Coefficients:
##                Estimate Std. Error z value Pr(>|z|)
## (Intercept)    1.152509   0.074332  15.505  < 2e-16 ***
## SpouseAlum     0.233712   0.098036   2.384 0.017129 *
## SportsAlum     0.151407   0.115283   1.313 0.189066
## UGAlumAwards  -0.187406   0.180344  -1.039 0.298732
## OtherUGAct     0.067163   0.086415   0.777 0.437030
## EverAssigned   0.499311   0.158570   3.149 0.001639 **
## BoardMember    0.686273   0.168011   4.085 4.41e-05 ***
## ChildAlum      0.097914   0.144314   0.678 0.497470
## ParentAlum    -0.007803   0.141300  -0.055 0.955962
## SiblingAlum    0.078758   0.106310   0.741 0.458796
## GradDegree     0.178044   0.124347   1.432 0.152192
## TotalReunions  0.066568   0.017340   3.839 0.000124 ***
## OnePlusEvents  0.259060   0.077225   3.355 0.000795 ***
## Years.Lapsed  -0.130215   0.003827 -34.027  < 2e-16 ***
## ---
## Signif. codes:  0 '***' 0.001 '**' 0.01 '*' 0.05 '.' 0.1 ' ' 1
##
## (Dispersion parameter for binomial family taken to be 1)
##
##     Null deviance: 9415.5  on 7119  degrees of freedom
## Residual deviance: 4866.7  on 7106  degrees of freedom
## AIC: 4894.7
##
## Number of Fisher Scoring iterations: 6
```

Fifth, Results Interpretations

The interpretations for std. error, z-values, and p-values are the same as those in a linear regression model. Below is a brief review of these concepts. For more details, please refer to Chap. 4.

Standard Error

It represents the uncertainty of the parameter estimate. Its value is closely related to the sample size. When the size of the data increases, or when the variations of the X variable increases, the standard error of the estimate decreases.

z-Value

It is the same as the *t*-value. It is calculated as the ratio of the parameter estimate to its standard error.

$$t = \frac{\hat{\beta}}{\text{StandardError}_{\hat{\beta}}}$$

This metric helps us to evaluate whether we are confident that the estimated value is either positive or negative, or statistically 0. As a rule of thumb,

- If the calculated *z*-value is far away from 0, specifically either larger than 2 (precisely 1.96) or smaller than −2, it indicates statistical significance. That means the parameter estimate is considered to be non-zero.
- If the calculated *z*-value is close to 0, in the range of [−2,2], it indicates statistical insignificance. That means, the parameter estimate is considered to be 0, or the corresponding *X*. The variable has no predictive power on whether the individual will donate or not (Fig. 6.3).

Pr(>|z|)

The *p*-value, when its value is very small, indicates that the parameter is non-zero.

Both *z*-value and Pr(>|z|) values tell us whether the parameter estimates are statistically significant or whether the estimates are essentially zero.

The estimated value:

- If positive, it indicates that the marginal impact of this variable is positive on the probability.
- If negative, it indicates that the marginal impact of this variable is negative on the probability.

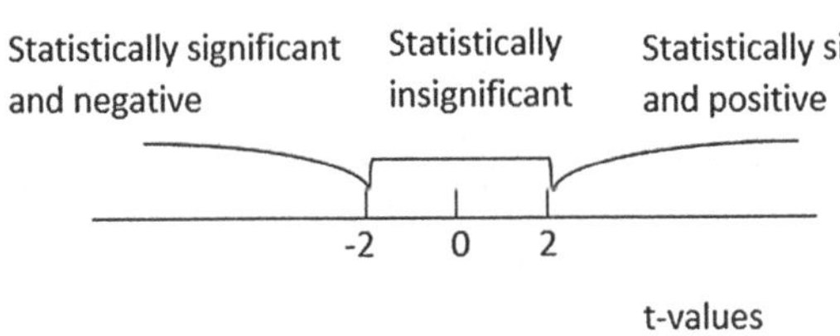

Fig. 6.3 Illustration of the implications of *z*-value (*t*-value)

Given the logistic transformation on the right-hand side of the model, the change in probability is not directly the parameter value, but rather it needs to be calculated. The calculation method is presented in details later in this chapter. This calculation is different from that in a linear regression, where the parameter estimates directly indicate the marginal impact of an *X* variable on the *Y* value.

Sixth, Try Different Model Specifications and Select the One That Works Best
In the model-building process, we need to try different model specifications by changing some of the X variables, for example, by adding or dropping variables, creating interactions, and creating new variables.

Then we can choose the best model, based on either

- in-sample fit metrics or
- out-of-sample fit metrics.

In most modeling exercises, the out-of-sample fit is the preferred metric for selecting the model winner; its details will be discussed in the next chapter.

Step 2: Calculate the Propensity Scores for Potential Targets

Now, applying the best model selected from Step 1, we predict the probability of donation from those who are invited to the grand reunion this year. These are alums who have graduated in multiples of 5 years (such as 5, 10, etc.) and are not necessarily included in the estimation data sample. These probability values are referred to as propensity scores.

Step 3: Find the Right Targets

Finally, in the third step, with the calculated propensity scores for each potential donor, we chose a threshold. We sent the promotional material to those with a propensity score above the threshold.

The End
The propensity score approach is widely used in marketing, particularly for selecting the right customers to target. It is not only prevalent in consumer marketing but also plays a crucial role in business-to-business (B2B) marketing.

In a B2B context, this approach helps sales teams determine which potential accounts (business customers) are worth pursuing. Here, the propensity score model is trained using account-level data and then applied to potential accounts to generate propensity scores. Sales representatives can then prioritize their efforts by focusing on accounts with the highest scores, thereby increasing the likelihood of a successful conversion.

In the preceding discussion, we explored the process of identifying target customers using propensity scores. Next, we will explore the underlying machine learning model behind this approach: the binary logit model.

Logistic Regression Model

The logistic regression model, as is implied by the name, is very similar to the regression model that we studied before, which requires a dependent variable *Y* and some independent variables *X*.

In the logistic regression model, the dependent variable is binary, taking values 0 and 1. Hence, it is also called the "binary logit" model. The goal is to examine how the probability of a "1" outcome changes with a set of explanatory variables *X*. Conceptually, this is similar to linear regression in that we relate an outcome to predictors, but logistic regression is designed specifically for binary decisions by modeling probabilities rather than the outcome directly.

Logistic Regression Versus Regression Model

Linear regression and logistic regression models are two widely used Machine Learning techniques, commonly applied across various fields, including Marketing. While they differ in several key aspects, they also share many similarities. Below, we outline the key differences and commonalities between them.

Key Differences

First, the Nature of the Dependent Variable

- In a logistic regression model, the *Y* variable takes only two possible values, 0 and 1, making it a type of **discrete model**. This model is not suitable when the *Y* variable can take more than two possible values.
- In a linear regression model, the *Y* variable is continuous and can take (theoretically) any real number values.

Second, the Range of the Predicted Values

- In a logistic regression model, the predicted outcome is a probability, which is constrained in the range of 0–1.
- In a linear regression, the predicted values are not constrained.

Third, Model Structure and Error Term Distribution

- In a logistic regression model, the predicted outcome is computed through a nonlinear transformation of the model parameters and X variables (details presented later in this chapter).
- In linear regression, the predicted outcome is computed through a linear computation of the model parameters and *X* variables.
- Due to the differences in the model structures, estimation techniques also differ. However, we do not need to worry about these technical details, as the programming language handles them automatically.

Fourth, Interpretations of the Parameter Estimates

- In a logistic regression model, due to the nonlinear transformation in calculating the outcome values, the parameter value cannot be directly understood as marginal effects of the X variables.
- In linear regression, the parameter estimates represent the marginal impact of the corresponding X variable on the Y value.

Key Similarities

First, the Selection Process of X Variables

- Both models use similar approaches for selecting explanatory variables, guided by the business context.
- Feature engineering: Both models follow the approach for creating and incorporating in the model dummy variables, interaction variables, and nonlinear transformations of the X variables. The process of creating new X variables based on the original data is also commonly referred to as "feature engineering."

Second, Statistical Inference

- The interpretation of standard errors, t-values, and p-values is essentially the same across both models. While there are minor theoretical differences, they are too technical and generally negligible for practical purposes.

Comparing the logistic regression model with the linear regression model helps build a better understanding of the logistic regression model. Next, we will explore the details of the logistic regression model.

Model Specifications

To set up the model, we connect the Y value with the observed X variables and the model parameters β, in a setting like

$$Y \sim \beta_0 + \beta_1 x_1 + \beta_2 x_2 + \ldots + \beta_k x_k + \epsilon$$

In regression, we change $\sim$ to =,

In the logistic regression model, we need to make two adjustments, including:

- Change the left-hand side to become a probability metric
- Change the right-hand side to ensure its possible values are bounded in the range of 0–1

After both adjustments, we obtain:

$$P(Y=1)=\frac{\exp(\beta_0+\beta_1x_1+\ldots+\beta_kx_k)}{1+\exp(\beta_0+\beta_1x_1+\ldots+\beta_kx_k)}=\frac{\exp(V)}{1+\exp(V)}$$

The second "=" is based on the definition below, for the purpose of simplifying notations,

$$V=\beta_0+\beta_1x_1+\beta_2x_2+\ldots+\beta_kx_k=\beta X$$

Based on this model, we can calculate another probability, which is the probability of not donating. Given that in the data, Y can only take two possible values: 1 for donating and 0 for not donating, we can calculate the probability of not donating as 1 − Prob(donating), that is

$$P(Y=0)=1-P(Y=1)=1-\frac{\exp(V)}{1+\exp(V)}=\frac{1}{1+\exp(V)}$$

The above three equations represent the logistic regression model; the complexity shows up on the right-hand side, where a nonlinear transformation is implemented after calculating the value of $V=\beta X$. This nonlinear transformation is also called the **logistic transformation**.

The logistic transformation ensures the right-hand side is always constrained in the range of 0–1, to match the value of the left-hand side $P(Y=1)$, given that probabilities are always in the range of 0–1.

Recall that in a linear regression model, the variable Y is on the left-hand side without any transformation, while the calculated value of βX is on the right-hand side, also without a logistic transformation. The key distinction between a binary logit model and a linear regression model lies in the use of the logistic function, also known as the logistic transformation. This difference is also reflected in how parameter estimates are interpreted, as discussed next.

Interpretation of the Model Parameters

In a linear regression model, such as

$$Y=\beta_0+\beta_1X_1+\beta_2X_2+\ldots+\epsilon$$

The parameter estimate for β_1 means that holding everything else constant, a unit increase in X_1 will lead to $\hat{\beta}_1$ units increase in Y.

In a logistic regression model, the change in a predictor variable, such as X_1, does not directly change the value of Y, since Y can only take the values of 0 or 1. Instead, it changes the probability of Y taking the value of 1, that is $P(Y=1)$.

The change in the probability is not directly related to the parameter value β_1, rather β_1 represents the change in the value of V. To determine how X_1 influences $P(Y = 1)$, we need to carry out the logistic transformation and calculate the corresponding change in probability.

For example, to understand the parameter estimate $\hat{\beta}_1$:

Denote the corresponding V value before the change in X_1 as V_0, and the V value after the as V_1. To change the X_1 value, we increase it by 1 unit, the value of V is increased by $\hat{\beta}_1$, that is

$$V_1 = V_0 + \hat{\beta}_1$$

The change in the probability can be calculated as:

$$\Delta P = P_1 - P_0 = \frac{\exp(V_1)}{1+\exp(V_1)} - \frac{\exp(V_0)}{1+\exp(V_0)}$$

Holding everything else constant, a unit change in X_1 leads to the change in probability as ΔP, as calculated in this equation. The resulting change in the value of P can be quite different from the value of $\hat{\beta}_1$; however, the direction of the change can be directly reflected by the sign of the estimated $\hat{\beta}_1$.

- If $\hat{\beta}_1 > 0$, a unit change in X_1 will **increase** the predicted probability, or the propensity score.
- If $\hat{\beta}_1 < 0$, a unit change in X_1 will **decrease** the predicted probability, or the propensity score.

Summary

This chapter reviews the concept of targeting and introduces one way of classifying different types of targeting strategies from the data analytics perspective.

This chapter introduces the process of targeting based on the **propensity scoring** approach, using the logistic regression model, which involves the following three steps:

Step 1. Build a logistic regression model, where the Y value indicates the success ($Y = 1$) or failure ($Y = 0$) for each customer.

$$P(Y=1) = \frac{\exp(X\beta)}{1+\exp(X\beta)}$$

In many ways, this model resembles a regression model, particularly in aspects such as:

- the setup of the *X* variables, such as different types of *X* variables, including
  - categorical variable
  - dummy variables
  - interaction variables

- the interpretations of the statistics, including
  - standard error of the parameter estimates
  - the *z*-values and *p*-values

In the meantime, it differs from a linear regression model in that the logistic regression model requires a nonlinear transformation—the logistic transformation:

- The range of the observed *Y* values are constrained to be binary, either 0 or 1; and the predicted values in the logistic regression models are probability values, and constrained to be in the range of 0–1. A linear regression model does not have those constraints in either the observed or predicted *Y* values.
- The interpretations of the parameter estimates are much more complex in the logistic regression model, as it requires calculations of the nonlinear transformation. Unlike in the linear regression model, the parameter estimates do not directly reflect the marginal impact of each *X* variable on the value of *Y*.

Step 2. Calculate the propensity score for the potential targets, based on the estimated model parameters.

Step 3: Select only those with a large enough propensity score as the actual targets.

This approach is applicable in many marketing situations when looking for the right customers to target, including both B2C and B2B.

Homework

1. Propensity score model

   Briefly describe the data needed to build a propensity score predicting the purchase probability of each potential customer, and the process to do so.
2. Logistic regression model

   When estimating a logistic regression model, or any statistical model, it's important to carefully examine the relationship between each explanatory variable (*X*) and the outcome variable (*Y*). In the example discussed in this chapter, one of the variables, **years_lapsed**, illustrates a common issue.

By definition, years_lapsed has a built-in relationship with the outcome: if $Y = 0$ (the individual has never donated), years_lapsed will tend to be large; if $Y = 1$ (the individual has donated), years_lapsed will typically be smaller. This negative association arises not from a meaningful predictive relationship but rather from how the variables are defined. In other words, observing a recent donation (i.e., a small years_lapsed value) doesn't necessarily indicate a higher likelihood of donating again; it simply reflects past behavior.

This spurious relationship often manifests in model output as either a very large positive t-value or a very small negative t-value, falsely suggesting strong predictive power. In such cases, the variable's inclusion may distort the model's interpretation and performance.

To address this, we should remove years_lapsed from the model and re-estimate the regression. Please proceed to drop this variable and report the updated model estimation results.

CHAPTER 7

Targeting and Propensity Score Model (Part II)

Learning Objectives

1. Implement the data conversion from text format to numerical values.
2. Articulate the challenges of working with unbalanced data.
3. Apply the SMOTE method to address the class imbalance issue.
4. Describe the difference between in-sample fit and out-of-sample fit in model evaluation.
5. Evaluate model performance using a confusion matrix and AUC value.

Introduction

This chapter builds on the previous one by addressing common challenges encountered when applying the propensity score approach to identify the correct targets.

The previous chapter introduced the propensity score method and the logistic regression statistical model. This chapter delves into additional considerations when analyzing real-world data, including

- How to redefine some text-based variables to make them numeric.
- How to deal with unbalanced data.
- How to compare and select the best model from a selection of models.

Supplementary Information The online version contains supplementary material available at https://doi.org/10.1007/978-3-032-11130-2_7.

© The Author(s), under exclusive license to Springer Nature Switzerland AG 2026

X. Dong, *Marketing Analytics and Data Science*,
https://doi.org/10.1007/978-3-032-11130-2_7

Initial Process of the Data

In this chapter, we utilize a publicly accessible dataset from a Portuguese banking institution. The dataset is archived at the UC Irvine Machine Learning Repository, and the detailed information can be accessed online.[1] It recorded a direct marketing campaign conducted by a Portuguese bank, and the demographic information of all the customers reached by the campaign, as well as the outcome of their decisions: whether to subscribe to a term deposit.

We use the smaller data file, "bank.csv", for demonstration purposes. One more thing to notice is that the file downloaded from the website is in a different format. The file downloaded from this course website is in CSV format and is ready to use by R or Python.

This book provides the R code needed to implement the analysis. The equivalent Python code is available online as supplementary material.

```
rm(list=ls())       # Remove everything in memory to start fresh.
set.seed(123)

# Import Data
bank_data <- read.csv("bank.csv", header = TRUE)
head(bank_data)

##   age          job marital education default balance housing loan  contact
day
## 1  30   unemployed married   primary      no    1787      no   no cellular
 19
## 2  33     services married secondary      no    4789     yes  yes cellular
 11
## 3  35   management  single  tertiary      no    1350     yes   no cellular
 16
## 4  30   management married  tertiary      no    1476     yes  yes  unknown
 3
## 5  59  blue-collar married secondary      no       0     yes   no  unknown
 5
## 6  35   management  single  tertiary      no     747      no   no cellular
 23
##   month duration campaign pdays previous poutcome  y
## 1   oct       79        1    -1        0  unknown no
## 2   may      220        1   339        4  failure no
## 3   apr      185        1   330        1  failure no
## 4   jun      199        4    -1        0  unknown no
## 5   may      226        1    -1        0  unknown no
## 6   feb      141        2   176        3  failure no
```

[1] https://archive.ics.uci.edu/dataset/222/bank+marketing

Recode the Variables to Be Numeric

The current data has many columns in text format; if we want to use them in the model, we need to convert them to numeric format (i.e., numbers). The following R code converts each of those variables and creates a new data frame, named `bank_data_use`

```
# first, remove some of the useless columns

bank_data_use <- subset(bank_data, select = -c(day, month, contact, poutcome)
)
```

For example, the `y` variable is defined as `yes` or `no`; we use the command `ifelse()` to change the values to be 1 and 0.

```
# change the y variable to be 1 for yes and 0 for no
bank_data_use$y <- ifelse(bank_data_use$y == "yes", 1, 0)
```

In using this command, `ifelse(equation, value1, value2),` , if the `equation` is true, the result is value1; otherwise, the result is 0.

We reassign these values to the variable bank_data_use$y, so that the new column replaces "yes" and "no" with 1 and 0, respectively.

Similarly, we can do this for many other variables, such as `job`. This variable takes many values, not just two possible values, unlike `y`. To determine the possible values for the variable "job", we use the "count" command from the "dplyr" package. This is a handy package for data analytics, which needs to be installed first.

```
#install.packages('dplyr')
library(dplyr)

##
## Attaching package: 'dplyr'

## The following objects are masked from 'package:stats':
##
##     filter, lag

## The following objects are masked from 'package:base':
##
##     intersect, setdiff, setequal, union

count(bank_data_use, job)

##              job   n
## 1         admin. 478
## 2    blue-collar 946
## 3   entrepreneur 168
## 4      housemaid 112
## 5     management 969
## 6        retired 230
## 7  self-employed 183
## 8       services 417
## 9        student  84
## 10    technician 768
## 11    unemployed 128
## 12       unknown  38
```

In fact, as shown above, it takes 12 different possible values, all of which are in text format. To incorporate such variables into the model, we can employ either of the two methods:

- Create a dummy variable for each value, which requires 12 columns to cover all the possible values.
- Create a factor variable so that the R program can automatically generate the 12 dummy variables and include only 11 of them, with one being the base variable.

To create the 12 dummy variables, we use the R command *ifelse()*.

```
# change job variable to be a few dummy variables
bank_data_use$admin   <- ifelse(bank_data$job=="admin.", 1, 0)
bank_data_use$bluecol <- ifelse(bank_data$job=="blue-collar", 1, 0)
bank_data_use$entrepr <- ifelse(bank_data$job=="entrepreneur", 1, 0)
bank_data_use$housemd <- ifelse(bank_data$job=="housemaid", 1, 0)
bank_data_use$managem <- ifelse(bank_data$job=="management", 1, 0)
bank_data_use$retired <- ifelse(bank_data$job=="retired", 1, 0)
bank_data_use$selfemp <- ifelse(bank_data$job=="self-employed", 1, 0)
bank_data_use$service <- ifelse(bank_data$job=="services", 1, 0)
bank_data_use$student <- ifelse(bank_data$job=="student", 1, 0)
bank_data_use$technic <- ifelse(bank_data$job=="technician", 1, 0)
bank_data_use$unemp   <- ifelse(bank_data$job=="unemployed", 1, 0)
bank_data_use$unknown <- ifelse(bank_data$job=="unknown", 1, 0)
```

Now, we can remove the original variable `job`.

```
bank_data_use$job  <- NULL
```

Similarly, we redefine the other variables.

```
# first, check the possible values of the variable "marital"
count(bank_data, marital)
##     marital    n
## 1 divorced  528
## 2  married 2797
## 3   single 1196
```

There are three possible values, allowing us to create three dummy variables. Alternatively, we could consider merging some of the values. In this case, given that the decision to purchase a financial product may be more influenced by marital status, and considering the small sample size for the "divorced" category, we can combine the values of "divorced" and "single". Therefore, we only need to create one dummy variable, indicating whether the person is married or not.

```
# change the married variable to be a dummy, married or not
bank_data_use$married <- ifelse(bank_data$marital == "married", 1, 0)
bank_data_use$marital <- NULL
```

Similarly, we can process the variable `education`. Instead of creating one dummy variable for each possible value of the education variable, we decided to create only one that indicates whether the person has a `tertiary`-level education (the highest level in the data).

```
# change the education variable to be dummy, tertiary, or not
count(bank_data, education)

##    education    n
## 1    primary  678
## 2 secondary 2306
## 3  tertiary 1350
## 4    unknown  187

bank_data_use$edu_ter   <- ifelse(bank_data$education == "tertiary", 1, 0)
bank_data_use$education <- NULL
```

Check the other variables.

```
# change the default, housing, and loan variables to be dummy
bank_data_use$default <- ifelse(bank_data_use$default == "yes", 1, 0)
bank_data_use$housing <- ifelse(bank_data_use$housing== "yes", 1, 0)
bank_data_use$loan    <- ifelse(bank_data_use$loan == "yes", 1, 0)
```

Now, check the structure of the new data file; all of the variables are numeric.

```
str(bank_data_use)

## 'data.frame':    4521 obs. of  24 variables:
##  $ age     : int  30 33 35 30 59 35 36 39 41 43 ...
##  $ default : num  0 0 0 0 0 0 0 0 0 0 ...
##  $ balance : int  1787 4789 1350 1476 0 747 307 147 221 -88 ...
##  $ housing : num  0 1 1 1 1 0 1 1 1 1 ...
##  $ loan    : num  0 1 0 1 0 0 0 0 0 1 ...
##  $ duration: int  79 220 185 199 226 141 341 151 57 313 ...
##  $ campaign: int  1 1 1 4 1 2 1 2 2 1 ...
##  $ pdays   : int  -1 339 330 -1 -1 176 330 -1 -1 147 ...
##  $ previous: int  0 4 1 0 0 3 2 0 0 2 ...
##  $ y       : num  0 0 0 0 0 0 0 0 0 0 ...
##  $ admin   : num  0 0 0 0 0 0 0 0 0 0 ...
##  $ bluecol : num  0 0 0 0 1 0 0 0 0 0 ...
##  $ entrepr : num  0 0 0 0 0 0 0 0 1 0 ...
##  $ housemd : num  0 0 0 0 0 0 0 0 0 0 ...
##  $ managem : num  0 0 1 1 0 1 0 0 0 0 ...
##  $ retired : num  0 0 0 0 0 0 0 0 0 0 ...
##  $ selfemp : num  0 0 0 0 0 0 1 0 0 0 ...
##  $ service : num  0 1 0 0 0 0 0 0 0 1 ...
##  $ student : num  0 0 0 0 0 0 0 0 0 0 ...
##  $ technic : num  0 0 0 0 0 0 0 1 0 0 ...
##  $ unemp   : num  1 0 0 0 0 0 0 0 0 0 ...
##  $ unknown : num  0 0 0 0 0 0 0 0 0 0 ...
##  $ married : num  1 1 0 1 1 0 1 1 1 1 ...
##  $ edu_ter : num  0 0 1 1 0 1 1 0 1 0 ...
```

After checking on the data types of all the variables, next we check on the values for the *Y* variable.

Unbalanced Data

When estimating a regression model, the outcome variable *Y* is typically continuous, meaning it can take on a wide range of values, theoretically from $-\infty$ to ∞. In contrast, a logistic regression model is used when *Y* is binary, taking only two possible values, 0 or 1.

The primary goal of the logistic regression model is to predict the probability that Y takes the value of 1, given a set of explanatory variables (predictors) X. In the context of machine learning, this is referred to as a classification problem. Depending on the nature of the task, various models can be employed to address classification problems; however, the logistic regression, also known as the binary logit model, is one of the most widely used approaches for binary classification.

Beyond binary classification, there are also multi-class classification problems, where the outcome can take more than two possible values. A typical example is recognizing handwritten digits, where the model must classify an input as one of the ten digits from 0 to 9. In this book, we will focus only on binary classification problems, where the outcome variable Y takes only two possible values (0 or 1). Multi-class classification is beyond the scope of this book.

What Is Unbalanced Data?

In the data sample used for model estimation (also referred to as the estimation sample or training dataset), if one outcome appears more frequently than the other, the training data is referred to as "unbalanced data".

In the bank data, this seems to be the case:

```
## # A tibble: 2 × 3
##       y     n freq_prop
##   <dbl> <int>     <dbl>
## 1     0  4000     0.885
## 2     1   521     0.115
```

In this dataset, there are only 521 positive cases where $Y = 1$, compared to 4000 negative cases where $Y = 0$. This means positive cases make up only 11.5% of the total observations, which is actually a typical pattern in marketing and many real-world datasets.

This class imbalance is frequently encountered in marketing analytics and predictive modeling. Some common examples include:

- Purchase Propensity Modeling: When predicting the likelihood of a customer subscribing to an online service, among those who visited the website the number of subscribers with $Y = 1$ is often far smaller than the number of non-subscribers ($Y = 0$).
- Click-Through Rate (CTR) Prediction: In online advertising, an important metric for evaluating an ad's performance is the ratio between the number of users who clicked on the ads over the total number of users

who saw the ads. Commonly, such a ratio is very low, indicating that among all those who saw the ads, only a tiny fraction (often less than 1%) of them actually clicked the ads ($Y = 1$), while the vast majority had no interactions ($Y = 0$) with the ads.

- Fraud Detection: In fraud detection models, the number of fraudulent accounts ($Y = 1$) is typically tiny compared to the large volume of legitimate accounts ($Y = 0$). If using such data to build a prediction model on fraud, it creates a significant challenge due to the imbalanced Y values.

Why Is This an Issue?

The best model we built matches the data when making predictions. Suppose one of the two classes appears significantly less often than the other. In that case, it is always safe for the model to avoid predicting cases that occur less frequently, thereby reducing prediction errors. As demonstrated above, the less frequent cases (such as user subscriptions ads clicking and fraud accounts detections) are more likely to be the ones that we are actually interested in.

For example, in this bank dataset, we hope to predict who is likely to purchase a Certificate of Deposit. The data shows that only 11.5% are "yes"; it is safer for the model to predict everyone to be "no", as the error would be pretty small. As a result, such a model would not be helpful for our business purposes.

How to Deal with the Unbalanced Data?

Before discussing the approach to deal with unbalanced data, we need to first distinguish between the training and testing data sets.

Training Versus Testing Data Sets

We split the data into two parts:

- Training Data (also called the estimation sample): we use these data to estimate the model parameters;
- Testing Data (also called the hold-out sample): we use these data to test the model built with the training data, by applying the estimated model parameters to the model and comparing the predicted values of Y with the actual values of Y in the data.

Size of the data when splitting:

- If we have a large set of data, it is common to split the data into two halves. Although we would like to keep a large training set to ensure we have enough data points for model estimation we need to keep in mind that the incremental computational cost of model estimation is much larger than that for model testing. This is not a big concern except when the dataset is really large, such as hundreds of mega bytes.

- If the data size is quite small, we need to ensure enough data is kept for training or model estimation, so it is common to have a smaller testing data set than the training data.

When splitting the data, the goal is to ensure these two data sets are essentially the same with no systematic differences. To achieve that, we need to pay attention to how the data set is organized.

- If the sequence of the data has no specific order, we can consider split the data in the middle. To do this, we need to ensure the condition is met. One way to check on that is after splitting the data, calculate basic statistics (such as mean, standard errors) for some each variable. Compare the statistics across the training and splitting sets and see if any obvious difference can be identified.
- If the data is ordered, we can randomly select some of the records to be training data and the rest to be testing data. For example, in a retail setting, customer records are often sorted by each customer's first purchase date. In a dataset ordered this way, the earlier portion contains customers with longer relationships (tenure) than the later portion. If you split the data into training and testing sets by simply taking the "first half" and "second half," the two sets will differ systematically, which can bias model evaluation. A better approach is to create the training set by randomly sampling customer indices. To do that, you can generate a set of random integers that meets two conditions: (1) the number of integers equals the desired training sample size, and (2) the integers range from 1 to the total number of observations. Use these integers as row indices to select records for the training set; assign the remaining records to the testing set.

There are no strict rules regarding the split ratio of the data; typically, we aim to ensure we have sufficient data for estimation. For example, the bank data contains 4521 records, and we select 2500 (slightly over half of the origi nal data) for training, reserving the remaining records for testing.
Method 1: Split the Data Close to the Middle

```
# total size of the data, number of rows
N <- nrow(bank_data_use) # number of observations in the whole data set

# row index for the training data
TrainSize <- 2500
idTrain <- 1:TrainSize

# row index for the testing data
idTest <- !(1:nrow(bank_data_use) %in% idTrain)

TrainData <- bank_data_use[idTrain,]
TestData  <- bank_data_use[idTest,]
```

Method 2: Randomly Sample the Data for Training

```
idTrain_random <- sample(1:N, size = TrainSize) #sampling training data from
the whole data set
idTest_random <- !(1:N %in% idTrain_random)
```

As we ask R to sample the records randomly, it may select a different set of samples each time this code is run. By using `set.seed(333)`, we ensure that the same sample is selected when the code is repeated.

For the analysis below, we adopt the results from the data split method.

Check the Similarities Between Training and Testing Data Sets

We need to make sure that the training and testing data sets are similar in their *X* and *Y* values.

First, check on the *Y* value.

```
print("Training Data")
## [1] "Training Data"
prop.table(table(TrainData[,"y"]))
##
##      0      1
## 0.8832 0.1168

print("Testing Data")

## [1] "Testing Data"

prop.table(table(TestData [,"y"]))

##
##         0         1
## 0.8866898 0.1133102
```

Based on the above calculations, the distribution of the *Y* values seem to be quite similar between the training and testing datasets.

To check on the *X* variables, we can use a new package `stargazer`, which provides a nice printout of the data summary statistics.

```
# install.packages("stargazer")
library(stargazer)

stargazer(TrainData,
          type = "text", digits=2,
          title="Summary Statistics of the Training Data")

##
## Summary Statistics of the Training Data
## ===============================================
## Statistic   N     Mean    St. Dev.   Min    Max
## -----------------------------------------------
## age       2,500  40.94    10.44      19     86
## default   2,500   0.02     0.13      0      1
## balance   2,500 1,379.97 2,665.86 -2,082 27,733
## housing   2,500   0.57     0.50      0      1
## loan      2,500   0.16     0.37      0      1
## duration  2,500  263.84   258.45     4    3,025
## campaign  2,500   2.78     3.00      1     32
## pdays     2,500  38.96    99.06     -1     871
## previous  2,500   0.53     1.60      0     20
## y         2,500   0.12     0.32      0      1
## admin     2,500   0.10     0.31      0      1
## bluecol   2,500   0.21     0.40      0      1
## entrepr   2,500   0.04     0.20      0      1
## housemd   2,500   0.02     0.15      0      1
```

```
## managem    2,500   0.22     0.41     0      1
## retired    2,500   0.05     0.21     0      1
## selfemp    2,500   0.04     0.20     0      1
## service    2,500   0.09     0.28     0      1
## student    2,500   0.02     0.14     0      1
## technic    2,500   0.17     0.38     0      1
## unemp      2,500   0.03     0.17     0      1
## unknown    2,500   0.01     0.09     0      1
## married    2,500   0.61     0.49     0      1
## edu_ter    2,500   0.30     0.46     0      1
## ---------------------------------------------------

stargazer(TestData,
          type="text", digits=2,
          title="Summary Statistics of the Testing Data")

##
## Summary Statistics of the Testing Data
## ===================================================
## Statistic    N      Mean    St. Dev.  Min    Max
## ---------------------------------------------------
## age        2,021 41.46      10.74     19     87
## default    2,021  0.02       0.13     0      1
## balance    2,021 1,475.46 3,386.90 -3,313 71,188
## housing    2,021  0.56       0.50     0      1
## loan       2,021  0.14       0.35     0      1
## duration   2,021 264.12    261.65     5    2,769
## campaign   2,021  2.82       3.24     1      50
## pdays      2,021 40.77     101.43    -1     804
## previous   2,021  0.56       1.80     0      25
## y          2,021  0.11       0.32     0      1
## admin      2,021  0.11       0.31     0      1
## bluecol    2,021  0.21       0.41     0      1
## entrepr    2,021  0.03       0.18     0      1
## housemd    2,021  0.03       0.17     0      1
## managem    2,021  0.21       0.41     0      1
## retired    2,021  0.05       0.23     0      1
## selfemp    2,021  0.04       0.19     0      1
## service    2,021  0.10       0.30     0      1
## student    2,021  0.02       0.13     0      1
## technic    2,021  0.17       0.37     0      1
## unemp      2,021  0.03       0.16     0      1
## unknown    2,021  0.01       0.09     0      1
## married    2,021  0.63       0.48     0      1
## edu_ter    2,021  0.29       0.46     0      1
## ---------------------------------------------------
```

The above two tables seem to be reasonably similar.

Alternatively, you can estimate the model multiple times, with resampled training and testing data each time. If the model fit metrics are dramatically different in one split, it means that in that trial, the training and testing data may have been quite different. It is worth investigating what drives that difference.

Balance the Training Data

When estimating (training) a model using an unbalanced training data set, the results will lead to problematic predictions. To tackle this problem, we need to balance the training data first before using it to estimate the model parameters.

There are multiple ways to balance the data:

- Reduce the size of the major class by randomly selecting only some data from this class. This method is also known as under-sampling. This is more appropriate when the total sample size is large enough.
- Increase the number of data points in the minor class. This method is also known as over-sampling.

For example, when the number of positive cases is only 10%, and the number of negative cases is 90%. We can choose *either* of the following:

- Under sampling: randomly select the negative cases from the original data, with a rate of 1/9; or
- Over sampling: randomly repeat the positive cases 9 times on average.

One common over-sampling approach is called SMOTE (Synthetic Minority Oversampling Technique). It creates multiple similar positive (minority) cases based on those in the data. We can apply this approach using the following R code, provided by the package `somtefamily`.

```
# install.packages("smotefamily")
library(smotefamily)

TrainData_bal <- SMOTE(TrainData,
                       TrainData$y,
                       K=3, dup_size = 0)$data
```

When using the function `SMOTE`, we need to make sure that:

1. The input data has only numeric values.
2. The output from `SMOTE` is fairly complicated, but the most useful one is the `$data` part.

Now compare the *y* values in the training data, before and after the SMOTE.

```
table(TrainData$y)

##
##    0    1
## 2208  292

table(TrainData_bal$class)

##
##    0    1
## 2208 2044
```

The number of negative cases ($Y = 0$) remains the same after applying SMOTE. The number of positive cases ($Y = 1$) increases dramatically, matching that of the negative cases. As a result, this new data set is much more balanced.

When comparing the training before and after running SMOTE, we notice that both the number of rows and columns are increased.

```
dim(TrainData)

## [1] 2500   24

dim(TrainData_bal)

## [1] 4252   25
```

After running SMOTE, the number of rows increased due to the oversampling of the positive cases. The number of columns is also increased, as the package created an additional column, called `class`, which is the `y` variable after the SMOTE process.

By now, we have completed the data preprocessing step and can estimate the logistic regression model using the same method discussed in the last chapter.

Estimate the Logistic Regression Model

Now we can estimate the logistic regression model using the training data after the resampling with SMOTE, using `glm()`.

After the SMOTE process, the new variable is labeled as "class" in the training data and is in numeric format. However, `glm()` needs the Y variable to be a factor, so we need to update that first.

```
TrainData_bal$y <- as.factor(TrainData_bal$class)
TrainData_bal$class <- NULL
```

Now, we are ready to execute `glm()`.

```
bal_result <- glm(y~., data=TrainData_bal, family = "binomial")
summary(bal_result)

##
## Call:
## glm(formula = y ~ ., family = "binomial", data = TrainData_bal)
##
## Coefficients: (1 not defined because of singularities)
##               Estimate Std. Error z value Pr(>|z|)
## (Intercept) -1.238e+00  4.694e-01  -2.638 0.008329 **
## age         -5.390e-04  4.903e-03  -0.110 0.912467
## default      3.993e-01  3.403e-01   1.173 0.240624
## balance      1.111e-05  1.607e-05   0.692 0.489053
## housing     -1.186e+00  9.299e-02 -12.757  < 2e-16 ***
## loan        -1.346e+00  1.518e-01  -8.867  < 2e-16 ***
## duration     5.478e-03  1.936e-04  28.293  < 2e-16 ***
## campaign    -1.438e-01  2.212e-02  -6.500 8.02e-11 ***
## pdays        1.476e-03  4.947e-04   2.984 0.002845 **
## previous     2.317e-01  3.270e-02   7.086 1.38e-12 ***
## admin        5.957e-01  4.247e-01   1.403 0.160692
## bluecol     -6.293e-01  4.217e-01  -1.492 0.135615
## entrepr     -9.772e-01  4.695e-01  -2.082 0.037381 *
## housemd     -7.063e-01  5.021e-01  -1.407 0.159466
## managem      7.022e-02  4.207e-01   0.167 0.867442
## retired      4.591e-01  4.442e-01   1.034 0.301313
## selfemp     -8.007e-01  4.689e-01  -1.708 0.087658 .
## service     -3.225e-01  4.365e-01  -0.739 0.459995
## student      9.148e-01  4.852e-01   1.885 0.059364 .
## technic     -2.526e-01  4.188e-01  -0.603 0.546319
## unemp       -5.415e-01  4.736e-01  -1.143 0.252923
## unknown             NA         NA      NA       NA
## married      8.467e-02  9.337e-02   0.907 0.364461
## edu_ter      4.101e-01  1.196e-01   3.429 0.000605 ***
## ---
## Signif. codes:  0 '***' 0.001 '**' 0.01 '*' 0.05 '.' 0.1 ' ' 1
##
## (Dispersion parameter for binomial family taken to be 1)
##
##     Null deviance: 5888.2  on 4251  degrees of freedom
## Residual deviance: 3954.9  on 4229  degrees of freedom
## AIC: 4000.9
##
## Number of Fisher Scoring iterations: 5
```

In this estimation result, the parameter estimates related to the dummy variable `'unknown job category'` are all `NA`. Because when we have 12 possible values in the `job`-related dummy variables, we will need to remove one of them as the base. In this case, the software automatically removed `unknown`, and parameter estimates for the other 11 dummy variables are all relative to the case of `unknown` for the `job` variable.

For example, the `student` variable is positive and statistically significant at a 0.05 significance level, indicating that a student is more likely to subscribe to a term deposit account compared to people whose job is unknown.

The above model includes all the *X* variables in the data. The results show that many of those variables included are not statistically significant. If we decide to include only some of the *X* variables, we can estimate another model. In the R code, we can specifically type out the variables to include:

```
bal_result2 <- glm(y~age+balance+loan+duration+campaign+pdays+previous+studen
t+edu_ter, data=TrainData_bal, family = "binomial")
summary(bal_result2)

##
## Call:
## glm(formula = y ~ age + balance + loan + duration + campaign +
##     pdays + previous + student + edu_ter, family = "binomial",
##     data = TrainData_bal)
##
## Coefficients:
##               Estimate Std. Error z value Pr(>|z|)
## (Intercept) -2.602e+00  1.900e-01 -13.695  < 2e-16 ***
## age          1.848e-02  3.804e-03   4.857 1.19e-06 ***
## balance      2.477e-05  1.505e-05   1.646   0.0998 .
## loan        -1.212e+00  1.420e-01  -8.537  < 2e-16 ***
## duration     4.954e-03  1.807e-04  27.424  < 2e-16 ***
## campaign    -1.452e-01  2.153e-02  -6.741 1.57e-11 ***
## pdays        5.552e-04  4.704e-04   1.180   0.2378
## previous     2.654e-01  3.330e-02   7.971 1.57e-15 ***
## student      1.793e+00  2.595e-01   6.910 4.83e-12 ***
## edu_ter      6.157e-01  8.656e-02   7.113 1.14e-12 ***
## ---
## Signif. codes:  0 '***' 0.001 '**' 0.01 '*' 0.05 '.' 0.1 ' ' 1
##
## (Dispersion parameter for binomial family taken to be 1)
##
##     Null deviance: 5888.2  on 4251  degrees of freedom
## Residual deviance: 4231.6  on 4242  degrees of freedom
## AIC: 4251.6
##
## Number of Fisher Scoring iterations: 5
```

So far, we have two different models with different features (*X* variables) in the model; how do we compare them and choose the best one?

Model Fit Metrics

In practice, we often try multiple models with different X variables. To compare how well each model fits the data and find the best-fit model, we need to develop some metrics. There are two types of model fit metrics, including in-sample fit and out-of-sample fit.

- In-sample fit refers to those metrics that use the estimation sample (training data) to evaluate and compare the estimated models.
- Out-of-sample fit refers to those metrics that use a different data sample (testing data) to evaluate the models.

Next, we'll discuss each type in detail.

In-Sample Fit

In estimating a linear regression model, we learned to check how well it fits the data using the metric R^2, which ranges from 0 to 1. In general, a model with a higher R^2 value indicates that the model explains a large share of the variation in the outcome variable and therefore is considered to fit the data better than a model with a lower R^2 value.

A similar metric for evaluating a logistic (binary logit) regression model is the model's log-likelihood value, which is widely used in practice. To understand it, it helps to start with the idea of a likelihood function.

The **likelihood function** is the joint probability of the observed data given the model parameters. It is important to realize that once the data are observed, the likelihood is viewed as a function of the parameters: it tells us how likely different parameter values are in generating the outcomes we actually saw in the data. Therefore, a logical way to estimate the parameters is to choose the values that maximize this joint probability. This is the maximum likelihood estimation method. All else equal, a model with a higher likelihood (i.e., it assigns higher probability to the observed data) fits the data better.

In practice, however, directly working with the likelihood can be numerically problematic. The joint probability is computed as the product of many individual probabilities, each between 0 and 1. With hundreds or thousands of observations in the data, this product can become extremely small, which creates numerical challenges and makes comparisons unstable. For example, if we have a data set with 10,000 observations, the likelihood value can be on the order of 10 to the power of –10,000, which can be too small to be accurately represented by a regular computer.

To avoid this, we use the **log-likelihood** instead. Taking logs converts products into sums: we take the natural log of each observation's probability and sum across all observations. Because $\ln(p)<0$, when $0<p<1$, the log-likelihood is typically negative, but it is much easier to compute and can be compared reliably. Since the log function is monotonic, the model with a higher log-likelihood also has a higher likelihood, we can compare models using log-likelihood values without changing the ordering.

The log-likelihood value from an estimated model can be obtained using the code:

```
logLik(bal_result)

## 'log Lik.' -1977.449 (df=23)

logLik(bal_result2)

## 'log Lik.' -2115.793 (df=10)
```

When comparing two models estimated using the same dataset, the model with a higher log-likelihood value is considered to provide a better fit to the data. Based on the reported values between the two models above, the first one appears to fit better.

In addition to the log-likelihood value of the estimated model, the output also includes the degrees of freedom (df = 23), which corresponds to the number of parameters estimated in the model. It's important to note that the log-likelihood value shares a similar feature with the R^2 metric in a linear regression model. Both of them increase with the number of parameters included in the model. This means that models with more variables will often appear to fit better, even if the improvement is due to overfitting rather than meaningful predictive gain.

Therefore, simply comparing log-likelihood values without accounting for model complexity can be misleading. In this case, the first model has a higher log-likelihood value, but it also includes more parameters (higher degrees of freedom). As a result, the comparison isn't entirely fair.

To understand the role of the number of parameters in model fit metrics, consider that a model aims to use a limited number of variables to capture the patterns in the data. When a model includes more parameters, it gains higher flexibility and can often achieve a better fit to the data. However, there is a limit to this benefit. In the extreme case where the number of parameters reaches the number of data points, such that each parameter fits a single point perfectly, the model achieves a perfect fit. However, such a model is useless, as it fails to reveal any meaningful patterns in the data and is useless for predictions. In practice, we seek models that not only fit the data reasonably well but also provide interpretable insights and reliable predictions. While increasing the number of parameters often improves the model's in-sample fit, it does not necessarily enhance its predictive power.

To address this, model selection criteria need to be adjusted with the number of parameters included in estimating the model. Two of the most popular metrics with these adjustments are the Akaike Information Criterion (AIC) and the Bayesian Information Criterion (BIC). These metrics penalize an excessive number of parameters, providing a more balanced approach for comparing models.

AIC can be calculated as:

$$\begin{aligned}\text{AIC} &= -2\times\ln(\text{likelihood}) + 2\times\text{DOF} \\ &= \text{Deviance} + 2\times\text{DOF}\end{aligned}$$

In this equation, DOF refers to Degree of Freedom, and is commonly evaluated by the number of parameters included in the model.

Note the −2 in the front; contrary to the log-likelihood value, the model with **lower** AIC fits the data better.

`glm()` model also prints out the Deviance metric, which is part of the AIC, without the DOF correction.

$$\text{Deviance} = -2\times\ln(\text{likelihood})$$

With the log-likelihood value (higher is better) and AIC value (lower is better), the Deviance is not always useful, even though the R package prints it out.

```
bal_result$aic

## [1] 4000.898

bal_result2$aic

## [1] 4251.586
```

According to the AIC values, the first model provides a better fit to the data.

The **Bayesian Information Criterion (BIC)** is quite similar to AIC, with some differences:

$$\text{BIC} = -2\times\ln(\text{likelihood}) + \ln(N)\times\text{DOF}$$

In the second term of this equation, N refers to the size of the data. In both AIC and BIC, the model with more parameters are panelized. The difference is the strength of the penalty. Compared to AIC, BIC penalizes additional parameters more strongly, especially when the sample size is bigger. Despite the differences between AIC and BIC, in most marketing analytics applications, these two metrics are consistent in selecting the model with the better fit.

Out-of-Sample Fit

Although adjusted R^2, AIC or BIC values are designed to penalize the number of parameters to reduce the possibility of selecting an unnecessarily complex model with too many parameters, they are still in-sample fit measures, because they are calcualted using the same data used for estimating the model. As a result, they are generally less convincing than out-of-sample model performance metrics, which evaluate how well the model predicts on new, unseen data (the testing data set).

In the above example, the model was estimated using only the training data. Now we can apply the model estimation results to the testing data and compare the predicted Y values from the model with those in the data.

```
# Prediction using the testing data
predTst1 = predict(bal_result, TestData, type="response")
predTst1[1:10]

##         2501         2502         2503         2504         2505         2
506
## 0.5476487245 0.1444595321 0.4118638724 0.4238610542 0.0000141647 0.0908666
593
##         2507         2508         2509         2510
## 0.2981723226 0.0536114747 0.2887914484 0.1040083734

predTst2 = predict(bal_result2, TestData, type="response")
predTst2[1:10]
##         2501         2502         2503         2504         2505         2
506
## 6.468666e-01 2.188159e-01 1.350936e-01 3.469088e-01 4.124828e-05 8.265987e
-02
##         2507         2508         2509         2510
## 1.545586e-01 1.455306e-01 3.316547e-01 2.440669e-01
```

The predicted values from the logistic regression model are probabilities, ranging between 0 and 1, while the observed Y values are binary, either 0 or 1. To evaluate the model's performance, we can compare the predicted probabilities with the actual outcomes. There are several methods to do this, depending on the evaluation criteria and the context of the analysis.

Method 1: Calculate the Log-Likelihood

In this approach, we evaluate how the model fits the data using the metrics that were optimized during model estimation: the log-likelihood value.

```
lnlike = sum(log(predTst1*TestData$y+(1-predTst1)*(1-TestData$y)))
lnlike

## [1] -980.4377
```

The model with the highest log-likelihood has the best out-of-sample fit.

Method 2: Create the Confusion Matrix

In the above method, the fit metrics are calculated based on the probability outcome directly from the model. When using a confusion matrix approach, we need to first translate the probability values into values of 1 and 0.

To do that, we first select a cutoff probability: if the predicted probability is above the cutoff, it is considered a positive prediction ($Y = 1$); if the predicted probability is below the cutoff, it is considered a negative prediction ($Y = 0$).

A common threshold to pick is 0.5:

```
thresh  <- 0.5              # threshold for categorizing predicted probabilitie
s
predicted <- cut(predTst1, breaks=c(-Inf, thresh, Inf),labels=c("0","1"))
table(TestData$y,predicted)
##    predicted
##        0    1
##   0 1461  331
##   1   62  167
```

The output table is referred to as the confusion matrix. In this table, the first column represents the number of cases predicted to be 0, also referred to as negative cases.

- Among them, 1461 are correct where the observed data is also 0. These are also referred to as “true negative” cases.
- Also in this column, 62 are incorrect where the observed data is 1. These are also referred to as “false negative” cases.

The second column represents the number of cases predicted to be 1, also referred to as positive cases.

- Among them, 331 are incorrect where the observed data is 0. These are also referred to as “false positive” cases.
- Also in this column, 167 are correct where the observed data is also 1. These are also referred to as the “true positive” cases.

The diagonal indicates the number of correct cases, and the off-diagonals list the number of incorrect instances.

Some Terminologies

Related to the confusion matrix, there is a long list of definitions. Among them, here are two of the most popular metrics:

Recall

Recall measures the model's ability to identify actual positive cases correctly. Specifically, it calculates the proportion of true positives that the model correctly predicts. Referring back to the confusion matrix in Table 7.1, this corresponds to the **second row**: among all cases where the true outcome (from the data) is $Y = 1$, how many were correctly predicted as positive?

To improve recall, we need to reduce the number of false negatives, referring to cases where the model incorrectly predicts $Y = 0$ when the actual value is $Y = 1$.

A practical example is COVID-19 testing. High recall is crucial because it ensures that individuals who are actually infected are correctly identified. Missing a positive case (a false negative) could delay treatment and increase the risk of spreading the virus. Therefore, improving recall is essential for timely intervention and public health safety.

Precision

Precision measures the accuracy of the model's positive predictions. Specifically, it calculates the proportion of predicted positive cases that are actually positive. Referring to the confusion matrix, this corresponds to the **second column**: among all cases where the model predicted $Y = 1$, how many are truly $Y = 1$?

To improve precision, we need to reduce the number of false positives, referring to cases where the model predicts a positive outcome ($Y = 1$) when the actual value is negative ($Y = 0$).

Back to the example of COVID-19 testing: High precision in a test means that when someone tests positive, they actually have the virus. This is important because a false positive result could lead to unnecessary isolation and disruption. For example, a person may need to be separated from their family or miss work, even if they are not infected with the virus. Improving precision minimizes these unnecessary consequences.

These two metrics are both critical, but they can be easily confused with each other. To help distinguish them, focus on the denominators in calculating these ratios:

- In recall, the denominator is the number of positive cases from the data.
- In precision, the denominator is the number of positive cases predicted from the model.

Table 7.1 Terminologies

	Predicted: 0	*Predicted: 1*
Data: 0	**True negative**	False positive
Data: 1	False negative	**True positive**

Both metrics are important as they help minimize the types of misclassification errors reflected in the confusion matrix. In an ideal model each should be close to 1. In practice, however, it is too costly to achieve high values on both metrics; some trade-off may be made, depending on the particular questions of the study. We may need to focus more on one of the two metrics.

Consider a common marketing application: identifing which customers should receive a coupon. To achieve this, we develop a model to predict who is likely to make a purchase. In this setting, missing those customers who are likely to make a purchase could result in a significant opportunity cost. Therefore, we aim to ensure that among the true buyers (i.e., the positive case in the data), nearly all of them received the coupon. This corresponds to a high recall value. At the same time, sending coupons to too many customers who are unlikely to make a purchase increases cost due to the producing, mailing and management of the coupons. To keep those cost under control, we also want a large share of the customers we send coupon to are true buyers, this corresponds to a high-precision value. When the cost of managing and distributing the coupons is significantly lower than the opportunity cost of missed sales, we can prioritize recall over precision, deciding on some extra coupon spending in order to avoid leaving true buyers outside of the coupon program.

In a different context, consider a lender building a model that predicts the payback probability of each applicant. In this case, the positive case can be defined as loans that are successfully repaid. If the lender is highly risk-averse and prefers to approve fewer loans in order to keep defaults to a minimum, the main concern is avoiding false positives, which are cases where the model predicts repayment but the borrower later defaults. In that situation, the lender should prioritize precision, choosing a model that ensures the approved loans are very likely to be repaid, even if that means turning away some applicants who would have repaid.

To put it simply:

- If the situation is "do not want to miss any true positive cases", prioritize recall.
- If the situation is "only trust those to be predicted to be positive cases", prioritize precision.

In cases where we want to increase both metrics and reduce both errors, and treat the different prediction errors with equal weights, a convenient way to calculate a summary of these two metrics is the *F*1 score. We then pick the model with the highest *F*1 score.

*F*1 Score

The *F*1 score evaluates the model based on both the above two metrics by calculating the harmonic average of the above two metrics.

$$F1\,\text{Score} = \frac{2}{\frac{1}{\text{Recall}} + \frac{1}{\text{Precision}}} = 2 \times \frac{\text{Recall} \times \text{Precision}}{\text{Recall} + \text{Precision}}$$

AUC Value and the ROC Curve

AUC value for the ROC curve is the most popular performance metric for classification problems.

AUC stands for "Area Under the Curve", which is not a particularly informative name. ROC stands for "Receiver Operating Characteristics", a similarly confusing name.

The term ROC originated in World War II when the military developed it to distinguish enemy aircraft from the noise in the sky (a binary classification problem). It is now widely accepted as a method for evaluating binary classification models.

Recall that the model predicts the probability of $Y = 1$. To convert these probabilities into either case ($Y = 1$ or $Y = 0$), we need to choose a threshold value. If the predicted probability is higher than the threshold, $Y = 1$; otherwise, $Y = 0$.

The ROC is plotted in a figure:

- True positive rate as the Y-axis
- False positive rate as the X-axis

The values plotted are obtained as the decision threshold decreases from 1 all the way to 0. When the threshold is 1, none of the cases will be classified as true cases, resulting in 0 true positive rate and 0 false positive rate; therefore, the curve starts at (0, 0). As the cutoff value decreases (not shown in the plot), more cases will be classified as $Y = 1$, the positive case.

Among these newly appointed positive cases, some are correct (true positives), while others are incorrect (false positives). If the model performs well, we will observe a faster increase in true positive rates compared to false positive cases. In that case, we observe a curve that rises sharply along the Y-axis, accompanied by a slight movement in the X-axis. The area under the curve (AUC) will be larger or closer to 1. Given that ROC curves can be difficult to compare directly across multiple models, we often summarize each curve with the AUC value. AUC captures the overall shape of the ROC curve in a single number, making model comparisons more straightforward.

If among the newly assigned positive cases, half are true and the other half are false, we know the model has no capability of distinguishing between the true and false positive cases, that is, applying the model is no better than having no-model. That ROC will be a 45-degree line on the plot. In this case, the area under the curve (AUC) will be close to 0.5. In other words, the lowest value of AUC is 0.5.

Therefore, the AUC is a convenient metric for evaluating the model's predictive power, as measured by the ROC. It has been widely adopted for model comparisons.

We can plot the curve representing the changes in the true versus false positive rates when the cutoff points are different. This curve is called the Receiver Operating Characteristics (ROC).

```
# install.packages("ROCR")
library(ROCR)

#For the first model, plot the ROC curve
pred1 <- prediction(predTst1,TestData$y)
perf1 <- performance(pred1,"tpr","fpr")
plot(perf1,col='blue')
```

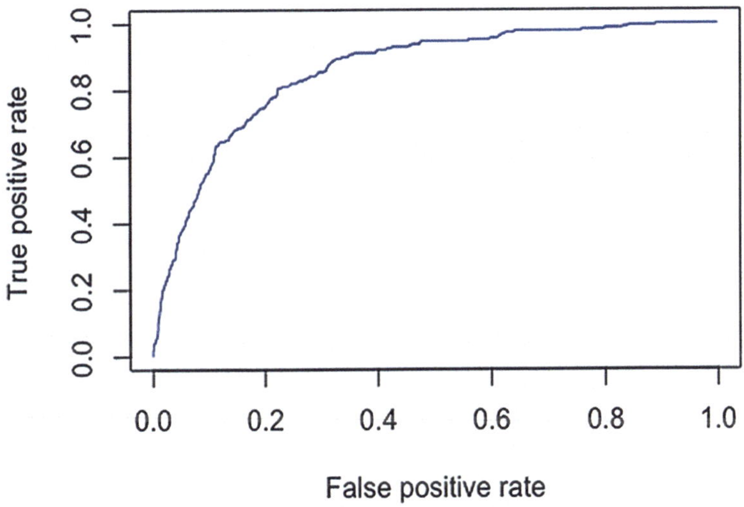


```
# calculate the AUC value
perf1 <- performance(pred1,measure="auc")
print(paste("AUC for the first model= ", perf1@y.values[[1]]))

## [1] "AUC for the first model=  0.856448358546473"

#For the second model, plot the ROC curve
pred2 <- prediction(predTst2,TestData$y)
perf2 <- performance(pred2,"tpr","fpr")
plot(perf2,col='green')
```

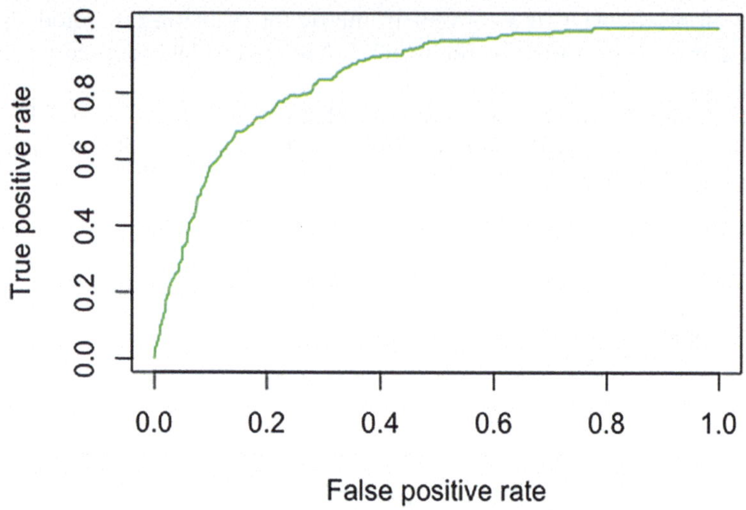


```
# calculate the AUC value
perf2 <- performance(pred2,measure="auc")
print(paste("AUC for the second model= ", perf2@y.values[[1]]))

## [1] "AUC for the second model=  0.854252768247035"
```

The code above shows how to plot the ROC curve and calculate the AUC value. To do that, we first need to install a new package called `ROCR`.

Regarding the two models above, the AUC values are remarkably similar to each other.

Summary

This chapter builds upon the previous one by providing explanations and the R code for addressing common real-world data challenges when constructing a classification model. In this chapter, we use a logistic regression model as an example; however, all these methods and concepts apply to all classification problems. Classification modeling is widely regarded as one of the most popular models in machine learning. The concepts discussed in this chapter include:

- Redefine some of the variables, and change from text to dummy variables.
- Split the data into training versus testing data.
- Before estimation, the training data can have the *Y* values balanced using approaches like SMOTE.
- Understand the difference between recall (denominator is true positive case in the data) and precision (denominator is the predicted positive case from the model).
- Compare models, the AUC metric, or the ROC curve obtained using the testing data.

Homework

The data can be downloaded from the textbook webpage, named "bank.csv".

Using this data file, build a propensity score model by following the steps:

1. Read in the data, clean up the data, and convert some variables from text format to dummy variables.
2. Split the data into training and testing data.
3. Balance the unbalanced data, using SMOTE.
4. Estimate three different models using the training data.
5. Compare the three models using the testing data, based on the AUC-ROC curve.

```
rm(list=ls())        # Remove anything that might currently be in memory so we
 can start fresh.
set.seed(123)
```

Step 1: Read in the Data and Convert to Numerical Data

Please read the data, and then remove these columns from the `data.frame`: day, month, contact, poutcome.

Recode the Y variable to be binary: using ifelse() to redefine the Y values from values of yes or no to numeric values of 1 and 0.

Choose some X variables that you believe are necessary to related with user's decision of whether to adopt the term project. Recode these X variables to dummy variables.

The X variables include

1. job
2. marital
3. education
4. default
5. housing
6. loan

Step 2: Split the Data into Training and Testing Data

Feel free to choose either one of the following two approaches when splitting the data:

- split the data at (or close to) the middle
- split the data randomly.

Answer the question: What is the size of your training data and testing data sets?

Step 3: Balance the Unbalanced Data

Please check and see whether the training data is balanced or not, by printing a frequency table of the values of *Y*.

Please apply the SMOTE approach to balance the data, for which you will have to install the package "smotefamily", using command `install.packages("smotefamily")`, followed by `library(smotefamily)`.

Now compare the balanced *Y* values with the unbalanced *Y* values, and comment on their differences.

Step 4: Estimate Two Models with Different Sets of X Variables on Training Data

Use the function `glm()`. Please refer to the reading or the notes for sample of R code. Note that please use only the training data not the whole data. Print out the estimation results for each of the two models, using function `summary()`

Step 5: Model Comparison Using ROC Curves and the AUC Values

To do this, you need to first install the "ROCR" package, using `install.packages("ROCR")`, followed by `library(ROCR)` to activate the package.

Then, your code should follow the steps, and repeat these steps for each model:

1. Predict the *Y* values using the testing data
2. Plot ROC curve and calculate the AUC value for each model

Finally, which model would you choose based on the AUC values?

CHAPTER 8

Forecasting and Bass Model

Learning Objectives

1. Apply the chain ratio method to generate forecasts in situations with limited or no historical data.
2. Explain the data-related challenges in forecasting problems.
3. Estimate the Bass model using regression techniques and produce forecasts using the Bass model parameters calculated from the regression results.
4. Explain the limitations of the Bass model and determine contexts in which it is appropriately applied.

Introduction

Sales forecasting is a critical challenge for nearly all businesses across many industries. The definition of "sales" can vary depending on the company and its priorities. For example, it might refer to product sales, service revenue, or even the number of new subscribers. Accurately forecasting future sales is essential but remains a complex task. Despite its challenges, effective sales forecasting is crucial for companies to plan strategically, guiding decisions on investments, resource allocation, business partnerships, and other key areas of focus.

As organizations increasingly recognize the value of their data, forecasting often becomes one of the first use cases they explore. Researchers have developed a variety of methods for sales forecasting, many of which fall under the domain of time series modeling. These models analyze data patterns over time to predict future outcomes. It is an in-depth subject in its own right and falls

Supplementary Information The online version contains supplementary material available at https://doi.org/10.1007/978-3-032-11130-2_8.

© The Author(s), under exclusive license to Springer Nature Switzerland AG 2026
X. Dong, *Marketing Analytics and Data Science*,
https://doi.org/10.1007/978-3-032-11130-2_8

beyond the scope of this book. Recently, machine learning techniques have also emerged as powerful tools for forecasting, focusing on identifying factors that drive fluctuations in data over time.

In this chapter, we explore common problems and solution approaches adopted in the context of marketing analytics. In particular, we begin by describing methods for making forecasts when data is unavailable. Next, we provide a detailed explanation of how to apply the Bass forecasting model to predict the adoption of new product categories. Finally, we introduce the Prophet model, a commonly adopted forecasting package in marketing practice.

Forecasting Methods

Although developing forecasting models is a key application of data analytics tools, forecasting tasks often face challenges due to limited data availability. In this chapter, we explore methods for both scenarios: when data is available and when it is not. We also discuss appropriate solutions for each situation. The following image provides an overview of various forecasting methods, though only some of them will be covered in this chapter (Fig. 8.1).

Forecasting with No Data

Sometimes, a company must predict a product's sales in a market before the product even exists. This situation often arises when making financial investments or marketing decisions. Despite the importance of such predictions, no historical data is available to build a machine learning model. In such cases, alternative methods, such as the chain ratio method or scenario analysis, can be employed to estimate future sales.

Chain Ratio Method

Any complex question can often be broken down into smaller, more manageable questions, with the final answer derived by combining the solutions to

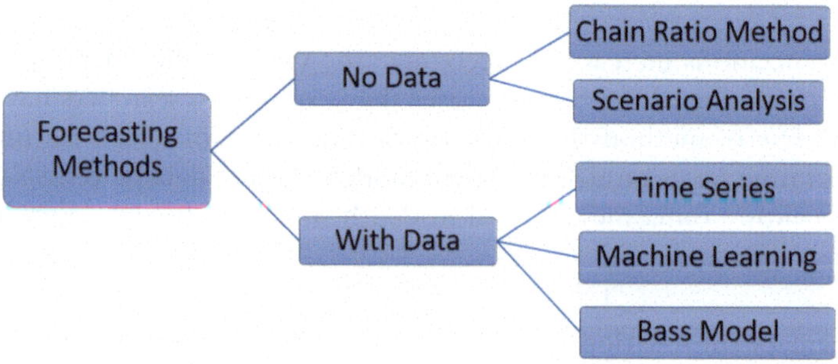

Fig. 8.1 A high-level overview of the general approaches to forecasting methods

these sub-questions. One popular approach for this is the chain ratio method, where the final result is obtained by multiplying the outcomes of a series of ratios, which are answers to a series of smaller questions.

Figure 8.2 illustrates the framework of the chain ratio method, which begins with an initial number and applies a sequence of ratios to it. Each of these components is multiplied together to arrive at the final answer. This approach breaks down a complex question into multiple smaller questions, making it easier to answer systematically.

The key advantage of this method is that, even when relevant data is unavailable, estimating individual ratios for each component is often more feasible than attempting to answer the original question outright. By decomposing the problem, we can make informed approximations and derive a reasonable estimate for the outcome.

Consider a real-world example: a pharmaceutical company was evaluating whether to renew its contract with a firm that provided a dedicated online social community for patients. During the discussion, the community provider showcased an impressive set of slides highlighting the benefits of their platform. The presentation was well-received, and most attendees were inclined to approve the contract renewal. However, a data scientist in the room posed a crucial question: How much additional revenue has this online community generated or is expected to generate? More importantly, does this additional revenue justify the cost of renewing the contract for another year?

She suggested starting by conducting a simple calculation on the spot at the meeting, without digging into the complex data process:

- Total number of members on the community platform (a number);
- Percentage of these members going to a doctor after hearing about a particular drug from the community that is produced by the pharmaceutical company (ratio 1);
- Percentage of these patents requesting the particular drug from their doctors (ratio 2);
- Percentage of doctors with these requests actually prescribe the drug requested (ratio 3);

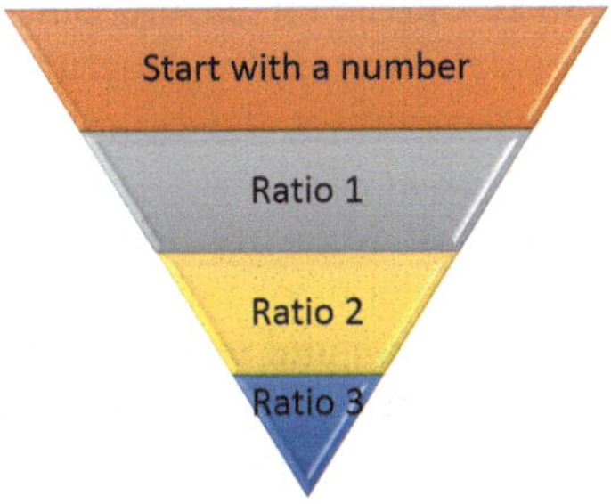

Fig. 8.2 Chain ratio method illustration

- The percentage of patients who actually fill the prescriptions by purchasing the drug (ratio 4).

With all these ratios, you can multiply them altogether and obtain a number that approximates the additional revenue generated by leveraging the online social community. Such a simple calculation reveals that the increased revenue is far from justifying the proposed renewal cost; therefore, the contract was not renewed.

Scenario Analysis
When predicting the future, we can consider multiple possible scenarios. A common approach is to define both a best-case and a worst-case scenario. For example, when a company is trying to launch a smart reusable water bottle that tracks hydration, the company can run the scenario analysis, by setting a best-case forecast of 200,000 units sold in the first quarter vs. a worst-case forecast of 40,000 units sold. Then the company can make decisions that are robust to both scenarios, in terms of the number of units to produce.

Forecasting with Some Data

When data is available, we can develop forecasting models to make informed predictions about future sales. Before diving into the details, it is important to distinguish between forecasting and prediction in the context of machine learning.

Machine learning excels at prediction, but not all predictive models are suitable for forecasting. Forecasting is a specific type of prediction that focuses on estimating future values based on historical patterns and trends. Forecasting is prediction with an explicit time structure. All forecasting is prediction, but not all prediction is forecasting. For example, as introduced in the previous chapter, the propensity score model is a prediction model but not a forecasting model. The propensity score model is commonly used to predict an individual's likelihood of making a purchase. It predicts probabilities for specific actions but does not forecast trends over time.

In this chapter, we focus on forecasting future sales to help companies allocate resources effectively. Since forecasting relies on time-dependent trends, sales data is typically structured as time series data, including time identifiers and sales metrics. To ensure accurate forecasting, it is essential to consider the following key factors:

Unit of Measurement: Sales metrics may be expressed in different units, such as millions or unit counts. Understanding the unit is crucial for accurate interpretation and model calibration.

Cumulative Versus Incremental Metrics: Distinguishing between these two types of metrics is crucial. A cumulative metric represents the total count over time (e.g., the total number of customers over a given period). An incremental metric reflects the change in count over a period (e.g., the number of newly acquired customers each month). Later in the chapter, the Bass model will be introduced; it is important to distinguish when to use the cumulative metric versus when to use the incremental metric during the model-building process.

Time Period Selection: Selecting the right time aggregation level is crucial for accurate forecasting. For instance, tracking daily new sign-ups may yield highly volatile data, making it challenging to generate reliable forecasts. Aggregating daily data into monthly figures smooths fluctuations, making trends more visible. However, this reduces the number of data points; for example, a one-year dataset contains 365 daily points but only 12 monthly data points. This trade-off between granularity and stability should be carefully evaluated when determining the data aggregation level before building the model.

In this chapter, we focus on two modeling approaches. The first one is a specialized model designed for forecasting new product sales: the Bass model. The Bass model is a fundamental tool in marketing analytics, closely aligned with the diffusion of innovation theory. It provides valuable insights into how new products or technologies are adopted over time, making it a powerful method for forecasting market trends. The second one is the Prophet model that can be applied to more general forecasting tasks than the Bass model.

Next, we introduce the diffusion of innovation theory first before diving into the details of the Bass model.

The Diffusion of Innovation Theory

The diffusion of innovations theory was first introduced by Everett M. Rogers in his 1962 book, Diffusion of Innovations. It explains how new ideas and technologies spread through a population over time.

According to this theory, the adoption of an innovation follows a predictable pattern. Initially, due to high uncertainty, only a small group of people is willing to adopt the innovation. As awareness grows and the benefits become more apparent, adoption gradually increases, reaching a larger audience.

Based on the timing of adoption, consumers can be classified into five categories (as shown in the figure below):

- Innovators (2.5%): Risk-takers who embrace new ideas first.
- Early Adopters (13.5%): Influential individuals who help drive broader adoption.
- Early Majority (34%): Pragmatic users who adopt once the benefits are evident.
- Late Majority (34%): Skeptical users who wait until most have adopted.
- Laggards (16%): Resistant to change, adopting only when necessary.

This classification enables businesses and policymakers to understand how innovations spread and strategize accordingly to accelerate their adoption (Fig. 8.3).

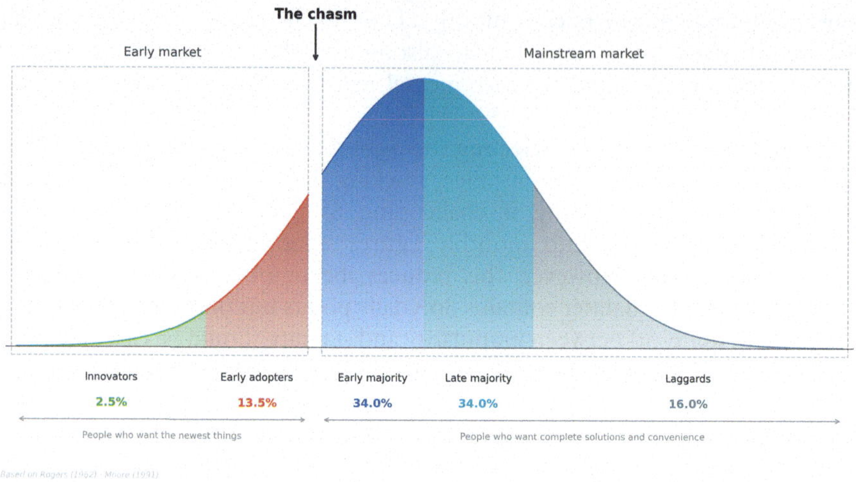

Fig. 8.3 The diffusion of innovation theory

These classifications are not scientifically rigid, but rather serve as a general framework for understanding adoption patterns. The distribution follows a bell-shaped curve, with probabilities of each group derived from the normal distribution function and measured by 1, 2, or 3 standard deviations away from the mean.

Bass Model

The Bass model is a straightforward model, developed based on many assumptions. When these assumptions are met or nearly met, the model's performance is remarkable, considering the simplicity of its setup. However, if the assumptions are not met, other, more sophisticated models are necessary to implement the sales forecasting task.

When building a sales forecasting model, the question it attempts to address is how many products will be sold in each time period in the future. Recognizing that this is quite difficult to achieve, the Bass model poses a slightly different question: at each time period, how many *customers* will adopt the new product? This transition from product-focused to customer-focused makes the sales forecasting task feasible.

The Bass model was developed to describe the diffusion process of a new technology by using a stylized model (Fig. 8.4).

Fig. 8.4 Professor Frank Bass (1926–2016)

A Brief History

The Bass model was first developed by Professor Frank Bass in 1966 and published in 1969 in a renowned journal *Management Science* (Bass, 1969). Fifty years later, in a comment paper published by Bass again in the same journal (Bass, 2004), Bass recounted the initial application of this model in forecasting the color TV sales in the late 1960s. According to the model forecasting, the color TV sales would peak in 1968 at 6.7 million units. Considering that color TV was extremely popular at the time, many Wall Street investors became angry with this forecast. As history shows, the sales actually did peak in 1968 at a slightly lower level than the model's forecast. However, the industry built a production capacity for 14 million color picture tubes, more than doubling the actual market sales.

Based on the historical background of the model and its success story, the Bass model has been proven useful not only in forecasting sales, as it was initially developed for, but also in assisting investment decisions.

Bass Model Setup

The goal of this model is to describe the process of how new products get adopted in a population, so that we can answer questions like:

- When will customers adopt a new product?
- How long before the sales of the new product reach its peak level?
- What will be the peak level sales?

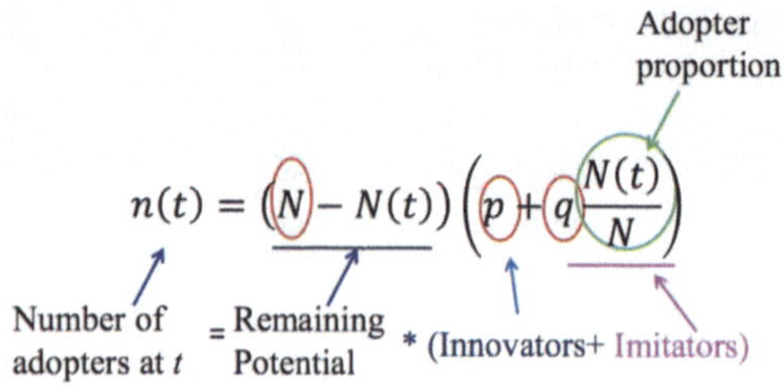

Fig. 8.5 Bass model formulation

In this model, instead of addressing "how many products are sold in each time period", it answers the question "how many customers adopt this new product in each time period"**.**

To do that, it assumes there are only two types of customers that exist to purchase the product in each time period. They are:

- Innovators: These are customers who purchase the product because other people **have not** purchased it. In this model, the proportion of innova-tors among all the customers who have **not** adopted the product is repre-sented by a parameter **p.**
- Imitators: These are the customers who purchase the product because other people **have** purchased it. The strength of this word-of-mouth effect is captured by a scale parameter **q.** (Figs. 8.5 and 8.6).

In the above model setup, the notations include:

- $n(t)$: sales in the time period t
- $N(t)$: total number of sales up to time period t (cumulative sales **up to** time period t) where $N(t) = n(0) + n(1) + \ldots + n(t-1)$
- p: coefficient of innovation, which is a ratio
- q: coefficient of imitation, which is not a ratio
- N: total size of the potential market

According to the model, if we know the values of the three parameters N, p, q, we will know the future sales at any given time.

To get an idea of the values for p and q, here is a table of these values for different products (adapted from Bass, 1969) (Fig. 8.7).

In the paper by Van den Bulte and Stremersch (2004), they surveyed an extensive list of durable products and suggested an average value of 0.03 for p and 0.42 for q.

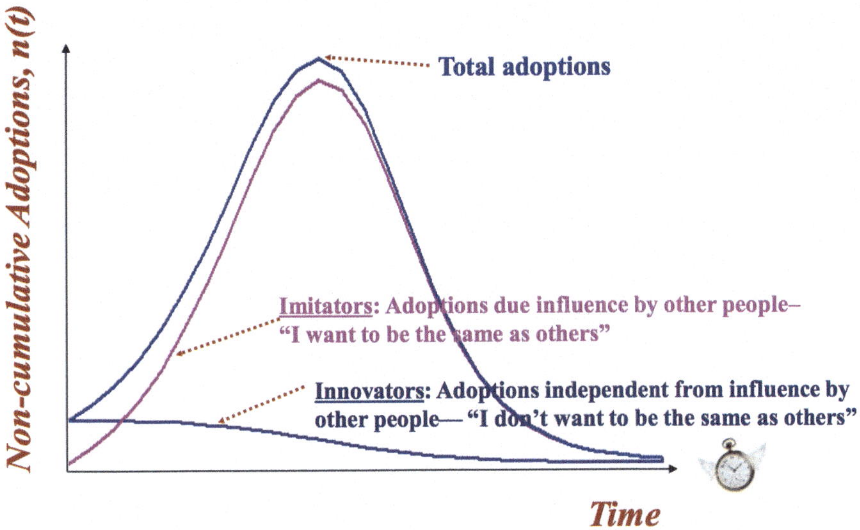

Fig. 8.6 Bass model illustration

Product	Period covered	N (1000)	p	q	R-squared
Electric refrigerators	1920-1940	40,001	0.002617	0.21566	.903
Home freezers	1946-1961	21,973	0.018119	0.17110	.742
Black and white TV	1946-1961	96,717	0.027877	0.25105	.576
Water softeners	1949-1961	5,793	0.017703	0.29695	.919
Room air conditioners	1946-1961	16,895	0.010399	0.41861	.911
Clothes dryers	1948-1961	15,092	0.017206	0.35688	.896
Power lawn-mowers	1948-1961	44,751	0.009184	0.33790	.932
Electric bed coverings	1949-1961	76,589	0.005876	0.25387	.976
Automatic coffee makers	1948-1961	58,838	0.017135	0.30145	.883
Steam irons	1949-1960	55,696	0.028632	0.32791	.828
Recover players	1946-1961	21,937	0.024796	0.65410	.899

Fig. 8.7 *N*, *p*, and *q* values for various products

Q: In this table, every row has $q > p$. What would happen if we had $q < p$?

Q: In this table, the last column shows *R*-square values; some are pretty high. Why?

Bass Model Estimation

Three parameters define the Bass model N, p, and q. Once these three parameters are known, the future sales can be forecasted. When historical data is available, these parameters can be estimated. In this section, we discuss the process of model estimation.

If you look at the table above, you may notice that the last column contains the R^2 value; does that ring a bell?

Regression! Exactly!

Now, let us explore how a regression model can be used to estimate the parameters of the Bass model.

Let us start with the model specification, that is

$$n(t) = \left(N - N(t)\right)\left(p + \frac{qN(t)}{N}\right)$$

Now, let us do some simple math: if we open up the two parentheses on the right-hand side, we get

$$n(t) = Np + (q - p)N(t) - \frac{qN(t)^2}{N}$$

With the historical data, we have the sales data for a product over a period of time. So, we know the values of $n(1)$, $n(2)$, …, $n(T)$, which is on the left-hand side, just like the Y variable in a regression context.

Recall that $N(t) = n(1) + \ldots + n(t-1)$, using this equation, we can calculate the values for $X_1 = N(t)$ and $X_2 = N(t)^2$, which are the X variables in the model. The regression analysis can be written as

$$Y = \beta_0 + \beta_1 X_1 + \beta_2 X_2$$

Comparing it with the above equation, we get

$$Y = n(t), X_1 = N(t), X_2 = N(t)^2 \tag{8.1}$$

Therefore, with historical data $n(1)$, $n(2)$, …, $n(T)$, we can get the Y variable, and calculate the two X variables. Then we can estimate the regression model and obtain the model estimates $\widehat{\beta_0}, \widehat{\beta_1}, \widehat{\beta_2}$.

Comparing the parameters from the regression model with those from the Bass model, we can get a system of equations with three equations to solve for three unknowns N, p, and q. They have unique solutions:

$$\begin{cases} \hat{\beta}_0 = Np \\ \hat{\beta}_1 = q - p \\ \hat{\beta}_2 = -q/N \end{cases} \Longrightarrow \begin{cases} N = \frac{-\hat{\beta}_1 - \sqrt{\hat{\beta}_1^2 - 4\hat{\beta}_0\hat{\beta}_2}}{2\hat{\beta}_2} \\ p = \frac{\hat{\beta}_0}{N} \\ q = -N\hat{\beta}_2 \end{cases} \quad 8.2$$

Using the above equations, we can compute Bass model parameters *N*, *p*, and *q* from the regression model results.

To summarize, the process of estimating the Bass model using historical data involves three steps:

First, using equations in Eq. (8.1) to get the *Y* and *X* variables.

Second, estimate the regression model with the *Y* values and two *X* variables.

Third, compute the Bass model parameters *N*, *p*, and *q* from the estimated values of the regression results, using Eq. (8.2).

The above demonstrates how to estimate the Bass model parameters using historical sales data. The idea is that, based on the Bass model and its assumptions, the diffusion process (sales over time) of a new product can be described by the three parameters. Therefore, as long as we have sufficient data to estimate the three parameters, we can describe the diffusion curve and, consequently, forecast future sales.

The future seems too easy to be true. However, past experiences have demonstrated the applicability of the Bass model. Although powerful, the Bass model can be applied only when its assumptions are met.

Assumptions and Limitations of the Bass Model

1. All three model parameters *N*, *p*, and *q* stay constant and do not change over time. This assumption can be violated, considering that the total potential market *N* could be reduced when a new technology was introduced to the market. Similarly, the word-of-mouth effect captured by the parameter *q* in the model could potentially vary over time.
2. Every customer in the market captured by *N* would eventually adopt the product.
3. Every customer purchases only the product once, with no repurchases. Once a customer has purchased the product, they are no longer in the market.

With only three parameters, the Bass model is remarkably simple, considering its capability to forecast the future. It achieves that power through the assumptions above. In other words, it applies only to some situations, as defined by the assumptions.

For example, some important considerations for product forecasting are missing in the Bass model, including product price and customer repurchase behaviors:

1. Price has been an important factor in a traditional sales forecasting model, but it is excluded in the Bass model. As a result, the model *cannot* be utilized to forecast the sales of a particular product in a competitive market, where the prices of the competing products play a critical role in each product's sales. Instead, the Bass model can be applied to forecast the sales of a specific product or a product category. That can also be seen from the table for the sample p, q values, where the first column lists product categories, rather than a product brand.
2. One of the key assumptions in the Bass model is that once a customer has purchased a product, they are out of the market. This suggests that the model is unable to account for repurchase behaviors. Strictly speaking, almost all products have customers who repurchase them, even big-ticket items like refrigerators. However, people usually purchase a new refrigerator many years after the previous one, so it is likely to be a different product from the first one, possibly due to technological changes. As a result, refrigerator forecasting can still apply the Bass model.

To address the important yet restrictive assumptions of the basic Bass model, its extension, a generalized Bass model, was developed and described in a 1994 paper published in *Marketing Science*, co-authored by Frank Bass, Triechy Krishnan, and Dipak Jain. In this generalized version, the model allows products to have an updated version and includes price in the model. The approach is beyond the scope of this book. For more details, please refer to the paper if you are interested.

With a good understanding of the Bass model and its limitations, next, we work with an example and illustrate how it can be applied using R coding.

An Example of Applying the Bass Model: EV Predictions

I have downloaded three data files for your class exercise and homework, which include:

- **EVMon.csv** and **EVQtr.csv**: Monthly and quarterly sales data on electric vehicles (EVs), from January 2012 to August 2017. EVMon.csv is for monthly sales; EVQtr.csv is for quarterly sales.
- **iPhone.csv** Quarterly sales data on iPhone, from the third quarter of 2007 to the fourth quarter of 2012.

Each of these data files has three columns,

- The first column identifies the year.
- The second column identifies the month (for monthly data) or the quarter (for quarterly data).
- The third column lists the number of units sold.

We use the EV monthly data as an example in the following notes, demonstrating the different methods in performing a forecasting analysis using the Bass model and the regression approach. The other two data sets are left for your exercise after class, with the iPhone data for your homework.

Estimate the Bass Model Using Regression

This book provides the R code needed to implement the analysis. The equivalent Python code is available online as supplementary material. Get started by using the monthly EV sales data file EVMon.csv.

```
rm(list=ls())
library(ggplot2)

EVData<-read.csv("EVMon.csv")

date_1 <- EVData[,1]
date_2 <- EVData[,2]
perSales <- EVData[,3]

startyear = date_1[1]
startmnth = date_2[1]

NTS = length(date_1)
```

Step 1. Get the Data Ready and Estimate the Regression Model, Based on the Bass Model

Based on the sales data, estimate the regression model

$$\text{sales}_t = \beta_0 + \beta_1 N(t) + \beta_2 N(t)^2$$

You can use the R command `cumsum()` to obtain the cumulative sum. But recall that

$$N(t) = n(0) + n(1) + \ldots + n(t-1)$$

`cumsum()` will add to the current t, but we need it to be added up to $t - 1$, so you need to adjust it. Also define $n(0) = 0$, using the following code.

```
# Need to remove the Last row, so that the n(t) is in the same row as N(0)+..
.+N(t-1)
# n(t) is the perSales in our code
# need to calculate the cumulative sales
NDATA <- length(perSales)
cumSales <- cumsum(perSales)

n0 <-0
Y <- perSales
X1 <- c(n0, cumSales[-NDATA])
X2 <- X1^2

reg.results=lm(Y~X1+X2)
summary(reg.results)

##
## Call:
## lm(formula = Y ~ X1 + X2)
##
## Residuals:
##    Min     1Q Median     3Q    Max
## -40101  -3254  -1457   5609  35112
##
## Coefficients:
##              Estimate Std. Error t value Pr(>|t|)
## (Intercept)  5.291e+03  2.141e+03   2.472  0.01607 *
## X1           6.131e-02  6.456e-03   9.496 6.69e-14 ***
## X2          -1.048e-08  3.169e-09  -3.308  0.00153 **
## ---
## Signif. codes:  0 '***' 0.001 '**' 0.01 '*' 0.05 '.' 0.1 ' ' 1
##
## Residual standard error: 10650 on 65 degrees of freedom
## Multiple R-squared:  0.8818, Adjusted R-squared:  0.8782
## F-statistic: 242.4 on 2 and 65 DF,  p-value: < 2.2e-16
```

Step 2. Recover the N, p, *and* q *Values*

Using the parameter estimates from the regression and applying them to the equations above, we can recover the N, p, and q values.

```
# Rename the parameter estimation results as a, b, and c, to simplify the equ
ation
  a <- reg.results$coef[1]
  b <- reg.results$coef[2]
  c <- reg.results$coef[3]

  # keep these equations unchanged, they are obtained from some linear algebr
a.
  N <- (-b - sqrt(b ** 2 - 4 * a * c)) / (2 * c)
  p <- a / N
  q <- b - p

  c(N=N,p=p,q=q)

##          N.X1 p.(Intercept)          q.X1
##  5.932345e+06  8.919292e-04  6.041717e-02
```

Step 3. Forecast Using the Bass Model

With the N, p, and q values, we can forecast the future sales of EVs by applying the Bass model.

```
bassModel <- function(p, q, N, initcase, T) {
  S = double(T)
  Y = double(T + 1)
  Y[1] = initcase
  for(t in 1:T) {
    S[t] = p * N + (q - p) * Y[t] - (q / N) * (Y[t] ** 2)
    Y[t+1] = Y[t]+S[t]
  }
  return(perSales = S)
}

predicted.perSales <- bassModel(p,q,N,n0,NTS)
predicted.cumSales <- cumsum(predicted.perSales)
```

Now using the following code to plot the results, with the actual and predicted values together:

```
# for monthly data, frequency is 12 for a year, setting up a time series form
at

# Use function ts() to create a time series format,
# for predicted sales
Spred <- ts(predicted.perSales, start = c(startyear, 1), frequency = 12)


# for actual sales
cases <- ts(perSales,           start = c(startyear, 1), frequency = 12)

# Then make the time series plots
ts.plot(cases, Spred, col = c("blue", "red"))
legend("topleft", legend = c("Actual Data", "Bass Model"),
                  fill = c("blue", "red"))
title(paste("Predicted cases","(p = ", format(p,digits=3),") ",
                              "(q = ", format(q,digits=3),") ",
                              "(N = ", format(N,digits=1),")"))
```

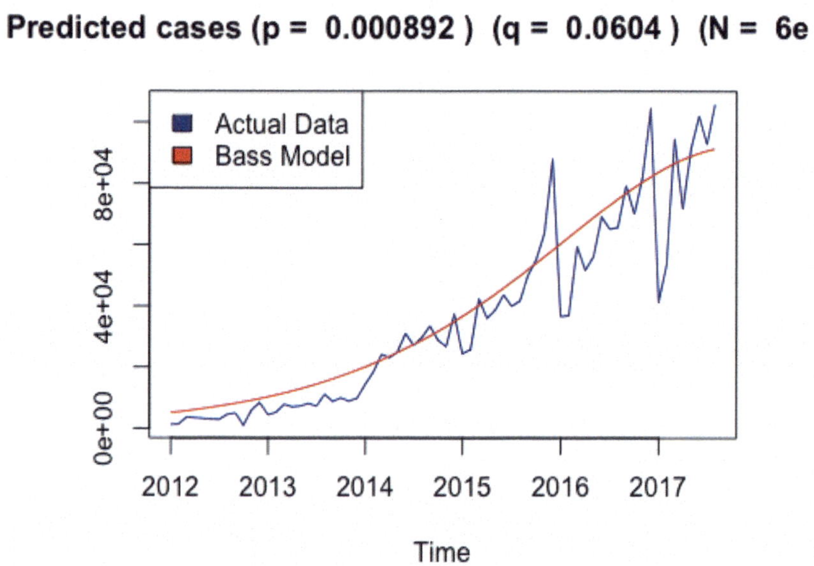


```
# for the cumulative sales numbers
cumSpred.ts <- ts(predicted.cumSales, start = c(startyear, 1), freq=12)
cumSales.ts <- ts(cumSales[-1],       start = c(startyear, 1), freq=12)

ts.plot(cumSales.ts, cumSpred.ts, col = c("blue", "red"))
legend("topleft", legend = c("Actual Cumulative cases",
                             "Cumulative from Bass Model"),
                  fill = c("blue", "red"))
title(paste("Predicted Cumulative cases","(p = ",format(p,digits=2),") ",
                                         "(q = ",format(q,digits=2),") ",
                                         "(N = ",format(N,digits=1),")"))
```

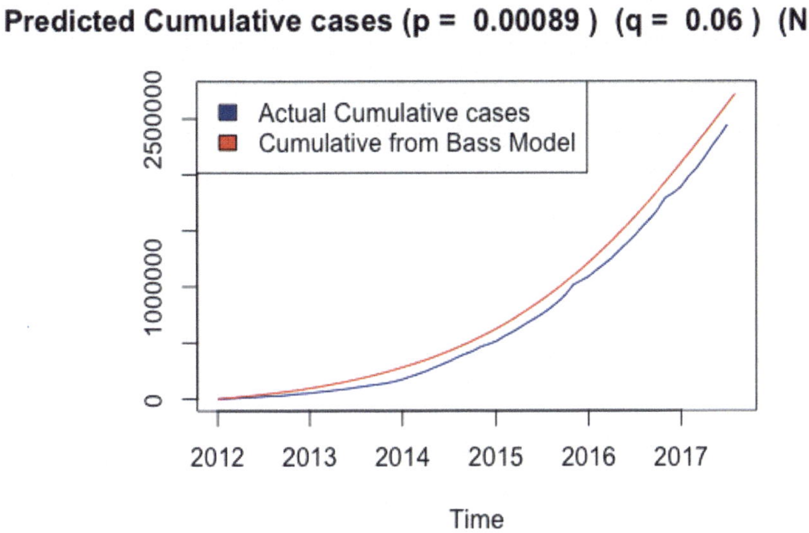

Estimate a Regression Model, with t *as* X *Variables*

In the previous estimation using a regression model, the *Y* values represent sales in each time period, while the *X* variables are derived based on the Bass model setup, including the cumulative sales up to the previous time period $(t - 1)$ and its square term.

Alternatively, we could estimate a regression model with only t values as the *X* variable. This approach is more consistent with the standard time series approach, where the sales trends are identified as a function of time. As we learned from the data plots, the diffusion curve is not linear over time, so we also added the square term t^2 as an *X* variable. The sales in each time period are the *Y* value. That is, we estimate the following model.

$$\text{sales}_t = \alpha_0 + \alpha_1 t + \alpha_2 t^2 + \varepsilon$$

For simplicity, you could define t to start from 1.

```
# Create the X variables in this regression, including t and t^2
Ndata <- length(perSales)
t<- seq(1,Ndata)
t2 <- t^2

# Estimate the regression model
reg.t<-lm(perSales~t+t2)
summary(reg.t)

##
## Call:
## lm(formula = perSales ~ t + t2)
##
## Residuals:
##    Min     1Q Median     3Q    Max
## -37719  -2883    183   3042  36320
##
## Coefficients:
##             Estimate Std. Error t value Pr(>|t|)
## (Intercept)   96.041   3786.038   0.025    0.980
## t            278.108    253.208   1.098    0.276
## t2            16.670      3.556   4.687 1.46e-05 ***
## ---
## Signif. codes:  0 '***' 0.001 '**' 0.01 '*' 0.05 '.' 0.1 ' ' 1
##
## Residual standard error: 10100 on 65 degrees of freedom
## Multiple R-squared:  0.8936, Adjusted R-squared:  0.8903
## F-statistic: 272.8 on 2 and 65 DF,  p-value: < 2.2e-16
```

Forecast the future by plotting the future cases together with the current cases.

```
T<-100

tt <- seq(1, T)
tt2 <- tt^2
futureX <- data.frame("t"=tt, "t2"=tt2)

predicted.tcases <- predict(reg.t, futureX)

# plot the predicted cases together with the actual data

Tpred <- ts(predicted.tcases, start = c(startyear, 1), freq=12)
cases <- ts(perSales,         start = c(startyear, 1), freq=12)
ts.plot(cases, Tpred, col = c("blue", "red"))
legend("topleft", legend = c("Actual Data", "Regression with t^2"),
                  fill = c("blue", "red"))
title(paste("Predicted cases Using Regression with t^2"))
```

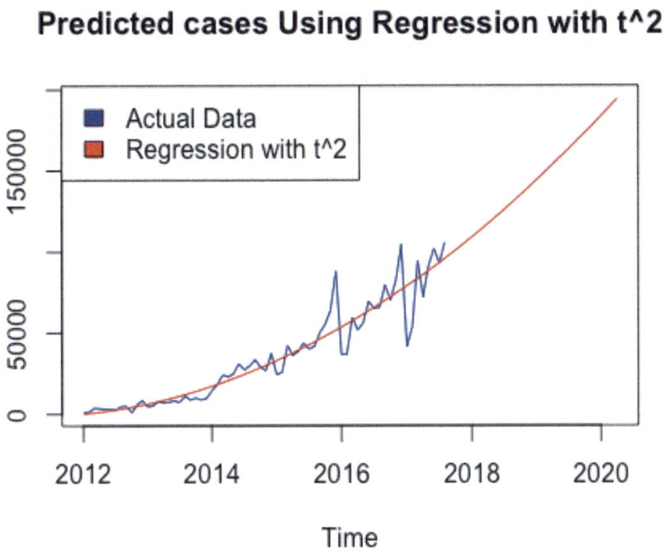

Based on the plot, the curve fitting is doing a nice job, especially at the early stage. The problem is extrapolation, or forecasting into the future. The plot shows that the sales in the future will grow much faster, because the estimated parameter for the t^2 term is positive and statistically significant. Comparing the two forecasts, although both are from a regression model, the second regression model using t and t^2 seems to be much more problematic, considering its forecast for the far future could exceed the total possible EVs on the market. The Bass model, on the other hand, offers a more realistic prediction for a future where the market's growth will be slowed over time.

The Prophet Model

As discussed earlier, the Bass model applies only to new products, as it is derived from the new product adoption process. The Prophet model is designed to forecast future sales based on past patterns and can be used to forecast sales of existing products.

Understanding the Prophet Model

The Prophet model is an open-source forecasting package developed by Meta's data science team. It has been widely adopted in practice due to its applicability across different forecasting problems and its accessibility to practitioners, who can apply it effectively without needing to engage deeply with its sophisticated statistical methods.

Its forecasting model is built upon an approach called the Bayesian structural time series framework. The details of the statistical model are beyond the scope of this book. It is helpful to know that this is a purely data-driven statistical approach for time series data, without employing any behavioral theories as utilized in developing the Bass model.

The Prophet model uses an additive model where the past time series data is decomposed into three components, including:

- Overall non-linear trends
- Seasonality
- Holiday effects

After deriving these three components from the historical time series data, the method extrapolates the data from the past, following these three components, and predicts the future.

Understanding the statistical basics of the Prophet model suggests that it is better suited for sales forecasting of relatively stable products, especially those with sales that are not strongly influenced by price changes.

An Example of Applying the Prophet Model

In this example, we obtained a dataset from Kaggle.com, a popular website for data competition. The link to the dataset is https://www.kaggle.com/datasets/matthieugimbert/french-bakery-daily-sales

The data is from a French Bakery, and it records the daily transaction details from January 1, 2021, to September 30, 2022. For illustration, the data were aggregated to the daily level before implementing the following R code.

Step 1: Get the Necessary Packages and Read in the Daily Data

```
# install.packages("prophet")
library(prophet)
library(dplyr)

bakery_data <- read.csv("Bakery daily Sales.csv")
head(bakery_data)
##         date dollar_sales Quantity
## 1 2021-01-02       987.85      581
## 2 2021-01-03      1014.30      564
## 3 2021-01-04       461.90      315
## 4 2021-01-05       515.70      309
## 5 2021-01-07       544.00      310
## 6 2021-01-08       533.60      316
```

The data contains three columns:

- The first column lists the date.
- The second column is the total sales in dollars each day.
- The third column is the total number of items sold each day.

In this example, we use only the first two columns to predict daily dollar sales.

Step 2: Get the Correct Variable Names

Before running the Prophet model, we must ensure that the variable names comply with the package's requirements. It requires the date variable named "ds" and the predictor variable named "*y*".

```
df_dollars <- bakery_data[,c("date","dollar_sales")] %>% rename(ds=date, y=do
llar_sales)
```

Step 3: Run the Prophet Model

Once the data is ready, we can call the function `prophet()`. In this setting, you can choose to specify the seasonality variables to be included:

- yearly.seasonality captures month-of-year patterns and holidays.
- weekly.seasonality captures day-of-week patterns.
- daily.seasonality captures hour-of-day patterns.

```
m <- prophet(df_dollars, yearly.seasonality=TRUE, weekly.seasonality = TRUE)

## Disabling daily seasonality. Run prophet with daily.seasonality=TRUE to ov
erride this.
```

Step 4: Predict the Future

Once the model is fitted, we can use it to predict the future. For example, predict the dollar sales in the next 300 days. The results can also be visualized with a plot.

```
future <- make_future_dataframe(m, periods = 300, freq = "day")
forecast <- predict(m, future)
plot(m, forecast)
```

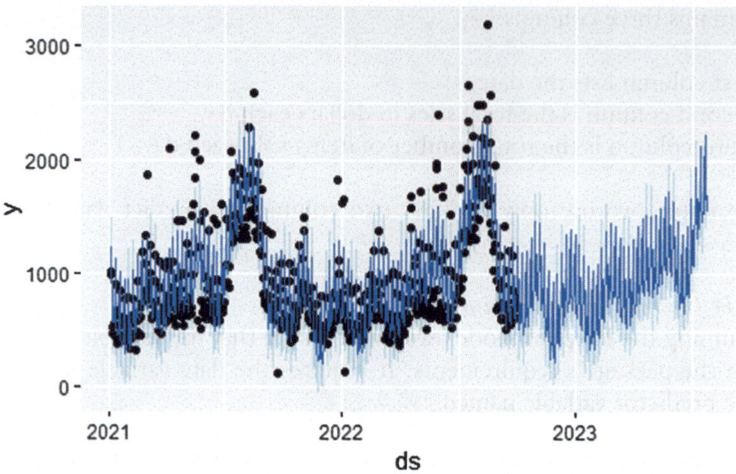

In the above plot, each black dot represents the observed data, and the blue line shows the predicted sales from the model. The fitted blue line reflects systematic variations over time, which are primarily explained by the seasonal components in the model.

To understand how the forecast is generated, the Prophet package also allows us to visualize the decomposed components obtained through its statistical analysis. In this case, the model breaks down the data from the past 600 days into multiple components, including the trend and seasonality. As specified in the code, two types of seasonality were included in this model. The daily seasonality is excluded, as time-of-the-day information is not included in the data. The three components can be plotted using the following command.

```
prophet_plot_components(m, forecast)
```

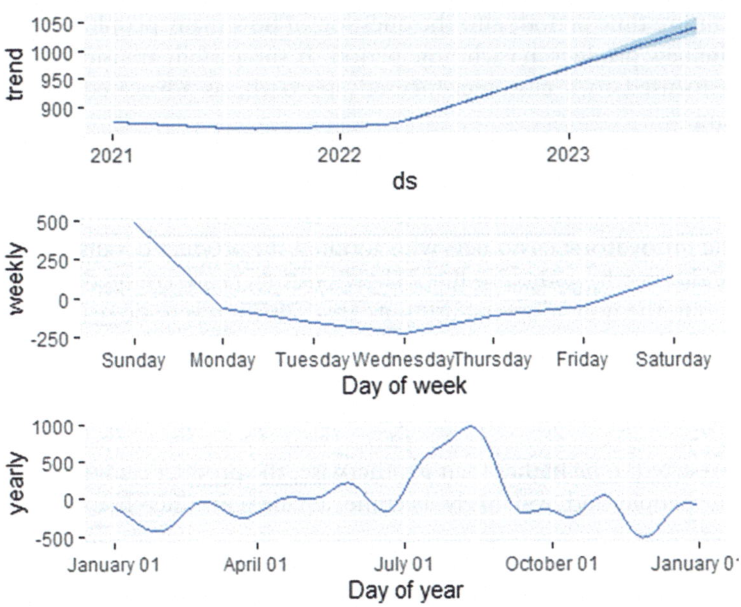

To understand the results from the visualization of the three components from the output:

- **The trend** is stable with a slight decrease until early in 2022, when the trend starts to increase. The shaded area toward the end of the plot indicates its increasing uncertainty for the later forecast.
- **The weekly seasonality** reflects higher sales on weekends, especially on Sunday, and a steady decrease into the middle of the week, until it starts to increase after Wednesday.
- **The yearly seasonality** shows elevated sales during the summer months, particularly in July and August.

Summary

Forecasting is a crucial yet challenging task. This chapter introduces three methods, each suited to a different type of forecasting task.

The first is the chain ratio method, which is used when no historical sales data are available. Although forecasting without data may seem impossible, it is a common situation in marketing. The chain ratio method addresses this challenge by decomposing the overall forecasting problem into a series of smaller, more manageable questions. Each question is tackled separately before combining the results from these multiple questions to produce the final answer.

The second is the Bass model, which is used to forecast future sales of a new product type. The Bass model is developed based on the diffusion theory in marketing, which translates the question of how many products to be sold in each future time period, into the question of how many people are going to adopt the new product in each time period. It successfully transitioned from a product-focused to a customer-focused approach, as advocated by modern marketing.

The Bass model identifies two distinct groups of customers and models them separately.

- The **innovators:** customers who adopted the product because other people did not have the product. These customers are counted by the ratio among the non-adopters, using p. This value tends to be really small.
- The **imitators:** customers who adopted the product because other people have adopted it. In this case, the more people adopted the product, the larger the number of imitators. It adds a scale q.

q is expected to be higher than p, otherwise, the product cannot be adopted by many people and will likely become a failed product before it crosses the chasm.

The Bass model is built upon several assumptions, with the key assumption being that customers purchase this product for one of two reasons: as an innovator or because other people have it. There is no consideration of price, promotion, or any marketing actions. A new product would be suitable in such a situation, but a particular brand in a competitive market would not.

Given that the Bass model is such a simple model, a common practice is that if we suspect the assumption may hold for the data, and may be aligned with the data generation process, we can always try to estimate the regression model to recover the N, p, and q values for the Bass model. If any of these values become negative, we know the Bass model does not fit. In that case, we need to work really hard to find a possible alternative for the forecasting task.

The third method is the Prophet model, which is suitable for products with relatively stable sales that are not highly sensitive to price changes. The Prophet model identifies long-term trends and seasonal patterns from historical sales data and then extrapolates such results into the future.

Forecasting plays a critical role in many business decisions, and the appropriate model depends on the business context. Each model serves different purposes, and it is essential to recognize their limitations in order to select the approach that best suits the situation.

Homework

1. Develop the Bass model using the data file iphone.csv.

   First, load in the data.

   Second, obtain the two X variables for the regression model.

$$\text{sales}_t = \alpha_0 + \alpha_1 N(t) + \alpha_2 N(t)^2$$

Remember $N(t)$ is calculated as the cumulative sales up to the time period $t - 1$.

Third, using the estimated values of $\widehat{\alpha_0}, \widehat{\alpha_1}$ and $\widehat{\alpha_2}$ to calculate the Bass model parameters N, p, and q. Comment on these values: what does each value mean in terms of the diffusion process of the smart phone? Based on these results, do you think the Bass model results are reasonable?

2. Forecast the future sales using a regression model with time t as X variable

$$\text{sales}_t = \alpha_0 + \alpha_1 t + \alpha_2 t^2$$

Based on the estimation results for the model parameters, plot the results, and comment on whether the forecasting makes sense or not.

References

1. Bass, Frank M. (1969), "A New Product Growth Model for Consumer Durables," Management Science, 15, 215–227.
2. Bass, Frank M. (2004), Comments on "A New Product Growth for Model Consumer Durables The Bass Model". Management Science 50(12_supplement):1833–1840.
3. Van den Bulte, Christophe and Stefan Stremersch, (2004), "Social Contagion and Income Heterogeneity in New Product Diffusion: A Meta-Analytic Test". Marketing Science 23(4):530–544.

CHAPTER 9

Bayes' Theorem and Marketing Attribution

Learning Objectives

1. Explain the concept of inverse probability and its role in statistical reasoning.
2. Interpret Bayes' theorem.
3. Describe the importance and influence of prior probability in Bayesian analysis.
4. Apply Bayes' theorem to address the marketing attribution problem.

Introduction

In this chapter, we introduce Bayes' theorem and lay the foundation of its application in solving two of the most important marketing analytics problems, which are explored in the next two chapters.

The first application addresses the marketing attribution problem, which aims to determine the contribution of each marketing investment to the business performance. The insight is critical for optimizing marketing budgets and future marketing investment decisions. Bayes' theorem is built upon the concept of inverse probability. It provides a natural framework for addressing this problem.

The second application involves effectively conducting online field experiments, one of the most reliable tools for evaluating marketing strategies in the digital environment. These experiments measure the effect of a marketing action by comparing outcomes between a treatment group, where the action is

Supplementary Information The online version contains supplementary material available at https://doi.org/10.1007/978-3-032-11130-2_9.

© The Author(s), under exclusive license to Springer Nature Switzerland AG 2026

X. Dong, *Marketing Analytics and Data Science*,
https://doi.org/10.1007/978-3-032-11130-2_9

implemented, and a control group, where it is not implemented. However, this approach inevitably incurs opportunity costs, as resources must be allocated to both groups before the results can be measured. The traditional statistical approach requires collecting large-enough samples of data from both groups, leading to a high opportunity cost. The Bayesian approach provides a more efficient solution by starting with smaller groups and continuously updating statistical inference as more data become available. This dynamic approach can dramatically reduce opportunity costs while achieving effective statistical inferences.

Bayesian statistics is named after Thomas Bayes, an English mathematician and theologian. However, while Bayes formulated the initial idea, much of the foundational development and popularization of Bayesian methods is credited to Pierre-Simon Laplace, a French mathematician from Normandy.

Thomas Bayes (1701–1761)

Pierre-Simon Laplace (1749–1827) (Both pictures are from Wikipedia.com)

Inverse Probability

In 1748, the Scottish philosopher David Hume published an essay, arguing that we cannot rely solely on traditional belief or testimony to infer cause and effect. Instead, he proposed that causal inferences must be grounded in experience, which we now call data, and that we should be content with fining only probable, rather than deterministic, causes. At a time when divine intervention was commonly accepted as the ultimate cause of all events, Hume's argument was both provocative and unsettling. Yet, it inspired many scholars to explore the nature of causality more deeply.

One such scholar was Thomas Bayes, an English clergyman and mathematician. In response to this philosophical challenge, Bayes devised a thought experiment to explore how probability could be used to infer unseen causes from observed data.

He imagined a scenario in which a ball is rolled onto a flat table from an unknown position. To estimate where the ball landed, he asked a colleague to toss pebbles onto the table and report whether each pebble landed to the left or right of the original ball. If, after three tosses, all pebbles landed to the right, Bayes reasoned that the ball was likely positioned toward the left side of the table. As more observations (data) are accumulated, his estimate of the ball's position becomes increasingly precise. Although he could never know the exact location with certainty, he was able to construct a probabilistic statement that improved with more data. This forms an informative illustration of the foundational concept for Bayesian inference.

To review this thought experiment:

- The "cause" is the ball landing at a specific position on the table.
- The "effect" is that the pebble will land either to the right or the left of the ball.

The central idea in this setting is: if we observe an outcome (the effect, the pebbles' relative positions), can we make a probabilistic statement about its underlying cause (the ball's position)?

This question shifts the focus from predicting outcomes based on known causes to inferring the most probable cause based on the observed effects, represented by data. In the eighteenth century, some hoped that such reasoning could be used to prove the existence of God. Bayes himself, however, was unable to achieve such a result, which may be one of the reasons that he never published his work during his lifetime.

This line of inquiry is the reverse of the typical causal question. Traditionally, we ask: "*Given a cause, what is the likely effect?*" But Bayes' approach asks the opposite: *"Given the effect, what is the likely cause?"* This formulation became known as the "inverse probability" problem.

For example, in everyday reasoning, we might ask, "*If I study hard, what is the probability that I will get an A on the exam?*" In contrast, inverse probability asks, *"Given that I got an A, what is the probability that I studied hard?"*

A more technical illustration involves diagnostic testing. For instance, *if someone tests positive for HIV, what is the probability that they actually have the disease?* Bayes' theorem provides a framework for answering such questions by combining prior beliefs with new evidence to update the likelihood of a cause.

For over two centuries, Bayes' rule was referred to as the "probability of causes" or simply inverse probability. It wasn't until the 1950s that the term

Bayesian became widely adopted, mainly due to the influence of the statistician R. A. Fisher, who was actually a leading critic of Bayesian methods. For more historical backgrounds and a fuller description of the Bayes method development and its applications during the World War II and in the decades that follow, please refer to the book, titled as "The Theory That Would Not Die—How Bayes' rule cracked the enigma code, hunted down Russian submarines and emerged triumphant from two centuries of controversy", authored by Sharon Bertsch McGrayne (2012).

Bayes' Theorem

Bayesian statistics is based on the well-known Bayes' theorem:

$$P(A|B) = \frac{P(B|A) \times P(A)}{P(B)}$$

Before explaining this equation, a couple of points to notice:

1. It relies on conditional probabilities.

Conditional probabilities are written in the form of $P(B|A)$, indicating that we know the information represented by A, we are trying to understand the probability of B happening, knowing that A has already happened.

For example, let B represent the event "I will arrive at school on time today", and let A represent the event "There is a traffic accident on the highway I take to school".

Without considering any additional information, we evaluate the unconditional probability $P(B)$, which reflects the general likelihood of arriving on time. However, once we learn that a traffic accident has occurred (event A), our assessment about B changes. The conditional probability $P(B|A)$ calculates the probability of arriving on time given that an accident has happened, which may be significantly lower than the unconditional probability. This illustrates (1) the importance of specifying what information we know in conditional probabilities, and (2) how new information can substantially alter the probabilities through conditional reasoning.

2. A and B switch positions in the Bayes equation

The switching of A and B is a key to making the "inverse probability" possible. When we want to calculate the inverse probability, Bayes' theorem instructs us on how to rely on what we know.

In the above example, the problem would be different. Now we observed B: a student arrived at school on time, and we are trying to infer A: the probability that an accident actually happened on their way to school. This illustrates the application of Bayes theorem in inferring the inverse probability and finding the cause based on the results.

Understanding Bayes' Theorem

The HIV test provides a valuable example for illustrating the components of Bayes' theorem within a cause-and-effect framework. In this context:

- Cause: The individual is actually infected with HIV.
- Effect: The medical test result is positive.

Bayes' theorem allows us to calculate the probability of the cause (actual infection) given the observed effect (a positive test result), by combining prior knowledge about infection rates with the accuracy of the test.

In this context,

- *B* refers to an outcome that we **observed** in the data. In this example, it relates to the observed effect: the medical test came out to be positive.
- *A* refers to a possible **reason** that is relevant to the outcome. In this example, it relates to the cause: the person is actually infected with HIV.

The question in this example is: given the test is positive (the observed effect), what is the probability that the patient is actually infected with HIV (a probabilistic statement of the cause)? Based on the notation defined above, the question can be written as calculating the conditional probability $P(A\text{-}actually\ infected | B\text{-}tested\ positive)$.

In this calculation, we are interested in the inverse probability problem: given that the medical test result came out as positive (effect), what's the chance that the person has the virus (cause)? This inverse probability can be easily confused with the probability in the original sequence: from cause to effect.

A common misperception is that if someone tests positive for HIV, they must be infected with the virus, especially if the test is considered accurate. However, this reflects a misunderstanding between forward (cause-effect) probability and reverse (inverse) probability.

The cause-and-effect probability refers to the likelihood that a test will yield a positive result when a person is actually infected. For example, a sensitivity of 98% means the test correctly identifies 98% of infected individuals. This gives the impression that the test is highly reliable.

However, the inverse probability, which refers to the likelihood that a person is actually infected *given* that they tested positive, is a different question altogether. Its value can be significantly different from the forward probability. To correctly calculate this inverse probability, we need to apply Bayes' theorem.

On the right-hand side of the Bayes equation,

- $P(B|A)$: Given a person is actually infected, the probability that the test will detect it and the test result will be positive.

This refers to the accuracy of the medical test. This is the cause-and-effect that we are more familiar with, and is what we know from the data (existing information).

- $P(A)$: The probability of any person who is infected with HIV, without conducting any tests on that person.

As we are trying to infer the probability of being infected given the test result, we need to formulate a "guess" of this probability *before* collecting information (data) regarding the test result. So, this is called the **prior probability**.

- $P(B)$: The probability of any person getting a positive result in the HIV test.

To have a positive test result, there could be two possible relevant scenarios:

1. The person has HIV, called A, and in the meantime, his test result came out to be positive, represented as $P(B|A)$.
2. The person does not have HIV, call it not A; however, his test result came out to be positive, represented as $P(B|\text{not } A)$.

To calculate the total probability of B, we can use

$$P(B) = P(B|A)P(A) + P(B|\text{not } A)P(\text{not } A)$$

On the left-hand side of the equation:

- $P(A|B)$: Given the test result, the probability of this person being actually infected with HIV is.

This is the outcome we are trying to obtain, also known as the **posterior probability**.

One way to comprehend Bayes' Theorem is to view it as a tool for calculating inverse probability:

Given an observed outcome, what is the probability that it was caused by a particular underlying reason?

To do this, Bayes' theorem combines two essential pieces of information:

- Prior Probability
  It provides an initial estimate of how likely the particular cause was the reason *before* seeing any new data. This probability reflects our baseline belief, often based on historical data or previous experiences.
- New Evidence
  Evidence is collected through observed data that updates our beliefs. For example, whether a medical test result is positive or negative. Keep in mind that the test results are the effect in the cause-and-effect paradigm. The question is to find the cause in this structure.

By combining prior beliefs with new information, Bayes' theorem allows us to form a **posterior probability**. This reflects a revised estimate from the prior probability and describes the likelihood of the particular cause of the outcome. The difference between the prior and the posterior probabilities is that the latter is informed by the observed outcome, while the prior is based only on past knowledge and experience.

Where to Find Prior Probability

One of the most transformative contributions of Bayes' theorem to statistical inference is the concept of prior probability. Unlike traditional (frequentist) approaches that rely solely on observed data, Bayesian inference enables the integration of relevant knowledge that exists outside the current dataset.

According to Bayes' theorem, the prior probability represents our belief about a parameter or hypothesis *before* observing the data. This belief can be informed by a wide range of sources, including:

- Expert knowledge or experience that is not explicitly recorded in the current dataset
- Findings from other analyses or studies that inform our understanding of the phenomenon of interest

Below are two examples of using prior probability in practice.

Example 1: Price elasticity in coffee sales

When analyzing how price affects coffee sales in a grocery store, we can use a linear regression model with sales as the dependent variable and price as one of the independent variables. Based on economic theory and experience, we expect the price coefficient to be negative (i.e., as price increases, sales decrease). However, due to data or modeling issues, we may occasionally obtain a positive coefficient for price.

In traditional modeling, analysts often respond by re-estimating the model or adjusting variables until the coefficient appears to be "correct". This informal process reflects the influence of external knowledge, which is not in the dataset, on the modeling decisions. The Bayesian framework offers a formal alternative: this prior belief about the direction of the price effect can be incorporated explicitly into the model through a prior distribution for the price parameter.

Example 2: Incorporating field knowledge in propensity modeling for the sales team

When building a propensity score model to predict whether a prospective customer will convert to a paid account, we typically use features from internal databases. However, a salesperson who has interacted directly with the prospect may possess valuable insights that are not captured in the recorded data. Such insights may include helpful information about the potential business customers, such as their business situations, intentions, or pain points. These insights can be translated into prior beliefs and incorporated into the model through the Bayesian approach, enriching the prediction process.

The Bayesian approach provides a formal method for incorporating additional information not directly recorded in the data to enhance the estimation of the posterior probability.

It is essential to note that prior probabilities must be based on external information. They cannot be derived from summary statistics or patterns in the current dataset. This distinguishes the prior from quantities like sample means or correlations, which are purely data-driven. In Bayesian analysis, the prior serves as a structured way to incorporate external knowledge into statistical models, enabling more informed and flexible inference.

A Numerical Example

Let's return to the HIV testing example and walk through it with actual numbers for illustration. Suppose the HIV test is highly accurate in detecting the virus when it is present, with a sensitivity of 98%. This means that if a person is infected, the test will correctly return a positive result 98% of the time:

$$P(\text{Positive}|\text{Infected}) = 0.98$$

At the same time, the test is also very good at ruling out infection when the person is not infected, with a specificity of 99%. That is, if the person is not infected, the test will correctly return a negative result 99% of the time:

$$P(\text{Negative}|\text{Not Infected}) = 0.99$$

(Please note: these numbers are used for educational purposes and do not reflect the actual performance of current HIV tests.)

Now here's the key question: If a person receives a positive test result, what is the probability that they are actually infected?

Most people would assume the probability is very high, perhaps close to 98%, but that assumption is incorrect. This is a classic example of an inverse probability problem, and to solve it correctly, we need to apply Bayes' theorem.

Now, setting up the problem:

Define:

- A: the person is infected with HIV
- nA: the person is not infected with HIV
- B: the test result is positive

We are interested in calculating $P(A|B)$: the probability that this person is infected with HIV, given that his test result came out to be positive.

To calculate $P(A|B)$, we need the following three items:

- $P(B|\mathrm{A})$

This conditional probability can be obtained after switching the positions of A versus B. It can be interpreted as the probability that the test is positive given the person is infected with the virus:

$$P(B|\mathrm{A}) = 98\%$$

- $P(A)$, the prior for A

This probability refers to the same problem as the posterior, but before the test-related data is collected. It can be interpreted as the probability that any person is infected with HIV.

To obtain the prior, we can get that information from some public data sources. According to data published on the US Centers for Disease Control and Prevention, in 2022, the rate was about 0.44% in the USA (438.2 per 100,000 people https://gis.cdc.gov/grasp/nchhstpatlas/tables.html)

$$P(A) = 0.44\%$$

- $P(B)$, the total probability of getting B

There are multiple ways to achieve a positive test result, i.e., to be B. In this example, there are two possible ways:

1. If the person is infected with the virus $P(A)$, and the test came out to be positive $P(B|A)$
2. If the person is not infected with the virus $P(nA)$, despite this, the test came out to be positive $P(B|nA)$

$$P(B) = P(A) \times P(B|A) + P(nA) \times P(B|nA)$$

To calculate this total probability, we already know the values of the first two terms $P(A)$ and $P(B|A)$, and we also need to know the other two items.

$$P(nA) = 1 - P(A) = 1 - 0.0044 = 0.9956$$

To calculate $P(B|nA)$, we need to know

- $P(nB|nA)$, the specificity of the test

It represents the probability of negative results when the patient is not infected with the virus. This is another way of showing the accuracy of the medical test. It is given as $P(nB|nA) = 99\%$

$$P(B|nA) == 1 - P(nB|nA) = 1 - 0.99 = 0.01$$

Note that this equation is very important and is essentially the key to solving this problem. It illustrates that event B (tested positive) is not equivalent to A (with virus). Even when the test is quite accurate, there is still a non-zero probability that it can make a mistake: providing positive results when the person is not infected.

Now we can plug in these numbers into the Bayes equation.

$$P(A|B) = \frac{P(A) \times P(B|A)}{P(A) \times P(B|A) + P(nA) \times P(B|nA)} = \frac{0.0044 \times 0.98}{0.0044 \times 0.98 + 0.9956 \times 0.01}$$
$$\approx 0.3022 = 30.2\%$$

This numerical example shows that even if a test appears highly accurate, a positive result for a randomly selected person does not necessarily mean that the person has a high probability of being infected.

To improve that result, one possible way is to increasing the 98% sensitivity rate to 99%. The calculation becomes:

$$P(A|B) = \frac{0.0044 \times 0.99}{0.0044 \times 0.99 + 0.9956 \times 0.01} \approx 0.3044 = 30.4\%$$

The result is barely changed.

The question is why. Even when the test seems to be quite good, the probability of being infected is only about 30% with a positive test result.

The answer does not lie in the accuracy of the medical test, but rather, the answer lies in the other component included in the Bayes equation: the prior.

To see that, let's change the prior from $P(A) = 0.44\%$ in the current calculation to $P(A) = 0.1 = 10\%$, then we can get

$$P(A|B) = \frac{0.01 \times 0.98}{0.1 \times 0.98 + 0.9 \times 0.01} \approx 0.916 = 91.6\%$$

Now the answer becomes 91.6%, a dramatic increase from 30.4%.

This may be one of the reasons that, in practice, before conducting any medical tests, it generally requires a doctor's notes or diagnosis. That diagnosis provides prior information indicating a reasonable chance that the person is infected with the virus. If conducting a medical test on any random person, the prior probability that the person has a particular disease can be very low; in such cases, the test result can be misleading.

This example illustrates how Bayes' theorem can be utilized to solve an inverse probability problem. In fact, in everyday life, we may encounter similar issues when trying to infer what has happened based on what we observe.

One such application in marketing is the marketing attribution problem, as discussed below.

Marketing Attribution

Marketing attribution, together with marketing mix modeling, represents one of the most widely applied analytics challenges within marketing organizations. While marketing mix models (MMM) focus on estimating the marginal contribution of each marketing channel, their primary goal is to guide budget allocation decisions across channels. For instance, MMM helps answer questions such as: Should more budget be allocated to Google search ads or to Google display ads? It emphasizes differences across marketing channels to optimize spending efficiency.

Marketing attribution, by contrast, addresses a related but different question: given an observed increase in sales, what proportion of that performance improvement can be attributed to each specific marketing action? Marketing attribution aims to allocate credit for conversions or purchases back to the particular marketing efforts that influenced customer behavior.

Before the rise of big data and digital marketing, it was difficult to track which customers interacted with which advertisements. As a result, evaluations of marketing Return on Investment (ROI) were typically conducted at the channel level. For example, a common retrospective statement from the 1990s might be: "Last quarter, we spent more on TV advertising than on radio, and sales increased. It seems the ROI for TV was higher than for radio". These insights were somewhat data-driven but often heavily reliant on intuition and coarse aggregate correlations.

The digital marketing boom fundamentally changed this landscape. Today, marketers can track individual-level interactions with digital advertisements across channels, commonly referred to as "touches". In most cases, a customer makes a purchase only after multiple such interactions, such as viewing ads across different platforms and at various times. Consequently, a key challenge in modern marketing analytics is distributing credit across these multiple touches after a customer converts, a problem widely known as multi-touch attribution (MTA).

An Example of Marketing Attribution

Here is an example of a Marketing attribution problem, which will be used to illustrate the solution methods.

An online shoe retailer, call it EShoe, purchases ads through digital channels, including

- Google Keywords ads
- Facebook social media ads
- TikTok ads

EShoe can track customer activity across these digital advertising channels. Additionally, it tracks both on-site behavior (after clicking an ad and arriving at EShoe.com) and purchase data from its users. For example, consider the journey of a single customer, as recorded by EShoe's analytics system:

- Monday: The customer clicked on a Google ad, visited the EShoe website, and browsed for a while, but left without making a purchase.
- Tuesday: She clicked on a Facebook ad, returned to the website, and browsed again, but still did not make a purchase.
- Wednesday: She clicked on a TikTok ad, and returned once more, and this time made a purchase that resulted in $50 profit for EShoe.

This customer interacted with three different marketing channels before making a conversion. The question is: How should the value of the conversion be attributed to each of the three ad channels (Google, Facebook, TikTok) that contributed to her path to purchase?

This is the core of the multi-touch attribution (MTA) problem, which seeks to assign credit across multiple customer touchpoints that preceded a successful conversion.

The solution to such a problem offers valuable insights that are critical to the success of marketing decisions:

- First, it is crucial to understand what drives each customer's decision. Understanding how each customer makes decisions and the factors that influence their decisions can help firms design targeted and more effective marketing strategies.
- Second, it provides a detailed view of the return on investment in each channel at the individual customer level and can help inform marketing investment decisions.

There are two broad categories of solutions to this problem:

1. Rule-Based Solutions

These are simple, predefined approaches (e.g., last touch, first touch, linear attribution) that assign credit according to fixed rules. They are easy to implement but may not reflect the actual impact of each channel.

2. Model-Based Solutions

These use statistical or machine learning techniques (e.g., logistic regression, Shapley value, Markov chains) to estimate the marginal contribution of each channel based on observed data. They are more sophisticated and better suited for complex customer journeys, but also require more data and technical expertise.

Rule-Based Solutions

Without implementing any sophisticated modeling, marketers commonly apply a rule-based method to attribute success to different touches across channels.

Last Touch Solution

As the last touch before the purchase was TikTok ads, based on this rule, all the credit will be given to TikTok, that is.

Last touch attribution results:

- Google ads: 0%
- Facebook ads: 0%
- TikTok ads: 100%

First Touch Solution

Some marketers believe that it is the first touch that attracted the customer's attention to the particular product, so all the credit should be given to the first touch.

First touch attribution results:

- Google ads: 100%
- Facebook ads: 0%
- TikTok ads: 0%

Linear Solution

Some marketers believe all the touch points should get an even credit, evenly distributed.

Linear attribution results:

- Google ads: 33.3%
- Facebook ads: 33.3%
- TikTok ads: 33.4%

These approaches are relatively easy to implement, but can be easily compromised due to their reliance on strong assumptions.

Model-Based Solutions

Model-based attribution solutions offer a more rigorous and data-driven approach to addressing the complexities of multi-touch attribution. Unlike rule-based methods, which provide fixed, deterministic outcomes, model-based approaches rely on statistical modeling and inference to estimate the probabilistic impact of each marketing touchpoint. These models can quantify uncertainty, uncover interaction effects, and produce more nuanced insights into how marketing actions contribute to conversions.

In the broader field of marketing analytics, marketing attribution models and marketing mix models (MMM) are widely regarded as two of the most essential tools for guiding marketing investment decisions and evaluating campaign effectiveness. While both aim to inform resource allocation, they differ in focus and granularity:

- Marketing mix models operate at the aggregate level, measuring the marginal contribution of each marketing channel (e.g., TV, radio, digital) to overall sales or performance metrics.
- Marketing attribution models, by contrast, focus on the individual customer level, evaluating how specific marketing touchpoints across the customer journey influence conversion likelihood.

Together, these two modeling approaches provide a comprehensive view of marketing effectiveness, from high-level budget allocation to personalized customer engagement.

To implement the model-based attribution solution, we resort to Bayes' theorem. It involves two steps that are detailed below.

Build Propensity Score Model

To estimate the probability of purchase given each user's interaction with any channel, we begin by constructing a propensity score model using a binary logistic regression approach, as discussed in Chaps. 6 and 7.

To develop this model, we require a dataset that includes the following information:

- Each user's interaction history with the firm's advertisements across multiple marketing channels
- The outcome variable indicating whether or not the user completed a purchase

This data allows us to model the likelihood of purchase as a function of marketing exposure and user behavior.

Here is an example data obtained from http://www.kaggle.com, which records the above information.

This book provides the R code needed to implement the analysis. The equivalent Python code is available online as supplementary material.

```
data_df <- read.csv("ind_attribution.csv")
head(data_df)

##                       cookie conversion Facebook Instagram Online_Display
## 1 00000FkCnDfDDf0iC97iC703B          0        0         1              1
## 2 0000nACkD9nFkBBDECD3ki00E          0        0         0              0
## 3 0003EfE37E93D0BC03iBhBBhF          0        0         0              0
## 4 0007iiAiFh3ifoo9Ehn3ABB0F          0        1         1              0
## 5 000A9AfDohfiBAFB0FDf3kDEE          0        0         0              0
## 6 000ADEAB7fBin9fEoFFnAohi9          0        0         0              1
##   Online_Video Paid_Search days_since_first
## 1            0           0               25
## 2            0           1                8
## 3            0           1                3
## 4            0           0                9
## 5            1           0                5
## 6            0           1                4
```

Each column in the data file:

- cookie: a unique identifier for each user.
- conversion: takes the value of one if the user made a purchase, zero if not. In this dataset, with 2000 data points, 1000 of them made a purchase (with a value of 1) and the other 1000 didn't (with a value of 0).
- Facebook: 1 if clicked Facebook ads, 0 otherwise.
- Instagram: 1 if clicked on Instagram ads, 0 otherwise.
- Online_Display: 1 if clicked online display ads, 0 otherwise.
- Online_Video: 1 if clicked online video ads, 0 otherwise.
- Paid_Search: 1 if clicked paid search ads, 0 otherwise.
- days_since_first: number of days since the first interaction

```
# drop the first column before running the model
data_use <- data_df[,-c(1)]

# Then, estimate the logistic regression model, with conversion as the Y vari
able
result = glm(conversion~., data=data_use, family="binomial")
summary(result)

##
## Call:
## glm(formula = conversion ~ ., family = "binomial", data = data_use)
##
## Coefficients:
##                  Estimate Std. Error z value Pr(>|z|)
## (Intercept)     -0.292695   0.121847  -2.402  0.01630 *
## Facebook         0.362064   0.121130   2.989  0.00280 **
## Instagram        0.290526   0.117744   2.467  0.01361 *
## Online_Display   0.187538   0.126762   1.479  0.13902
## Online_Video     0.618796   0.130821   4.730 2.24e-06 ***
## Paid_Search      0.017584   0.110841   0.159  0.87395
## days_since_first -0.020182  0.006398  -3.154  0.00161 **
## ---
## Signif. codes:  0 '***' 0.001 '**' 0.01 '*' 0.05 '.' 0.1 ' ' 1
##
## (Dispersion parameter for binomial family taken to be 1)
##
##     Null deviance: 2772.6  on 1999  degrees of freedom
## Residual deviance: 2730.7  on 1993  degrees of freedom
## AIC: 2744.7
##
## Number of Fisher Scoring iterations: 4
```

With these model estimation results, we can calculate the probability for each person to convert, given their exposure to each marketing channel. These estimated probabilities will be used in the next step to attribute the success of a purchase to the contributing channels. Importantly, we only need to calculate these probabilities for the 1000 users who actually made a purchase, as attribution focuses on explaining observed conversions, and those who did not convert are excluded from this step.

In the code below,

- First select those who made a purchase in the data, conversion=1

```
library(dplyr)
# select the rows only with the conversion = 1, then drops the conversion col
umn
data_convt <- data_df %>%
  filter(conversion == 1) %>%
  select(-conversion)

channel_names <- c("Facebook", "Instagram", "Online_Display",
                   "Online_Video", "Paid_Search")
# count the number of channels
nchannels <- length(channel_names)
```

- Then calculate the five probabilities for each person

To achieve that, we conduct a loop, one for each channel. When calculating the probability for each channel, only the X variable corresponding to that particular channel takes a value of 1, while all other channels take a value of 0. All the probabilities are stored in the dataframe pred_prob.

```
# create a new data frame with the same number of rows as data_convt, and the
 number of columns as the number of channels

pred_prob <- data.frame(matrix(0, nrow = nrow(data_convt), ncol = nchannels))
colnames(pred_prob) <- channel_names

# Calculate the cause-effect probability for each channel, for each user
for (i in 1:nchannels) {
  data_convt[, channel_names[i]] <- 1
  data_convt[, channel_names[-i]] <- 0
  pred_prob[, i] <- predict(result, newdata = data_convt, type = "response")
}
head(pred_prob)

##    Facebook Instagram Online_Display Online_Video Paid_Search
## 1 0.4319949 0.4145336      0.3897786    0.4957540   0.3501918
## 2 0.4669368 0.4491800      0.4238539    0.5310316   0.3829783
## 3 0.4319949 0.4145336      0.3897786    0.4957540   0.3501918
## 4 0.5173353 0.4994577      0.4737349    0.5808104   0.4316528
## 5 0.5022061 0.4843267      0.4586691    0.5660017   0.4168652
## 6 0.4769955 0.4591857      0.4337397    0.5410695   0.3925603
```

Note that these probabilities do not need to add up to 1 for each individual (row).

```
colMeans(pred_prob)
##        Facebook       Instagram Online_Display    Online_Video     Paid_Search

##       0.4777217       0.4600054       0.4346867       0.5414413       0.3936925
```

With these cause-effect probabilities calculated, we can obtain the inverse probability using Bayes' theorem.

Attribution Using Bayes' Theorem

Before applying Bayes' theorem, we first need to determine the prior probability for each marketing channel.

As discussed earlier, the prior represents our initial belief about the likelihood of each channel contributing to a conversion, *before* observing any outcome data. If we were working within the company, this information would reflect actual strategic decisions and budget allocations, and we would be aware of these values. However, in this example, we do not have access to the company's marketing plan.

Instead, we construct the priors based on reasonable assumptions guided by the following logic:

- Facebook and online video are considered the most critical channels, likely receiving the highest investment. $P(F) = 0.3$, $P(OV) = 0.3$
- Instagram and online display are assumed to be less central, with lower investment. $P(I) = 0.1$, $P(OD) = 0.1$
- Paid search is assumed to fall somewhere in between. $P(S) = 0.2$

Note that these prior probabilities must sum to 1.

These assumptions enable us to assign relative prior probabilities to each channel, which we will utilize in the Bayes framework to estimate their contributions to conversion outcomes.

```
values <- c(.3, .1, .1, .3, .2)
prior_prob <- (matrix(values, nrow=1))
#colnames(prior_prob) <- channel_names
prior_prob

##      [,1] [,2] [,3] [,4] [,5]
## [1,]  0.3  0.1  0.1  0.3  0.2
```

To apply Bayes' theorem using R, we can either use for-loops to calculate the probabilities for each person and each channel, which may take a much longer time. Alternatively, we can leverage R's capability of calculating multiple items in a matrix simultaneously.

Given that in the Bayes equation, we need to calculate the prior× cause-effect probability for each channel, we can do this using straightforward R code.

```
pp <- sweep(pred_prob, 2, prior_prob, FUN="*")
```

In the above code, the `sweep` function allows the vector `prior_prob` to sweep over the matrix `pred_prob` across all rows. This is much faster than using for-loops.

To calculate the posterior probability for each channel for each person, we need to calculate the ratio of each number over the sum in the same row. To achieve that, we use the following R code.

```
denominators <- rowSums(pp)
posterior <- sweep(pp, 1, denominators, FUN = "/")
```

In this result, we can get the posterior probability of each channel for each individual.

We can obtain the average values across all the converted users.

```
colMeans(posterior)
##      Facebook      Instagram Online_Display   Online_Video    Paid_Search
##    0.30237442     0.09703396     0.09166681     0.34295773     0.16596708
```

Summary

Bayesian statistics opens up another door to addressing many data-driven problems. It differs from classical statistics in that it addresses the inverse probability problem by using Bayes' theorem. Although a simple equation, Bayes' theorem has profound implications and impacts on data statistics and interpretations.

The chapter presents two applications of Bayes' theorem in this module, centered around its definition of addressing the inverse probability issue.

1. From the medical test accuracy to the test results reading
2. From the propensity score modeling to the marketing attribution problem

In addition, the chapter offers a detailed discussion of the marketing attribution problem, along with its easy rule-based solutions, as well as the Bayesian approach. As one of the two most widely studied and discussed marketing analytics problems, marketing attribution has been extensively examined. The Bayesian approach has by far the most theoretical foundations.

Homework

1. **Cancer test**

The lung cancer incidence is about 50 cases per 100,000 people, so the prior probability that a randomly selected person has lung cancer is $P(\text{Dis}) = 50/100{,}000 = 0.05\%$.

(a) Suppose the sensitivity of the test is 99%, meaning the probability of testing positive given the disease $P(\text{Pos}|\text{Dis}) = 99\%$.
(b) The specificity of the test is 95%, meaning the probability of testing negative given no disease $P(\text{Neg}|\text{No Dis}) = 95\%$

If any random person is tested positive for lung cancer, what is the probability that they have the cancer? Calculate this posterior probability and discuss what drives the results to be surprisingly small.

2. **Terrorists detection**

In a city of 1 million residents, there are 100 terrorists and 999,900 non-terrorists. In an attempt to catch the terrorists, the city installs an alarm system with a surveillance camera and automatic facial recognition software. The software has two failure rates of 1%:

(a) The false negative rate: If the camera sees a terrorist, a bell will ring 99% of the time, and it will fail to ring 1% of the time.
(b) The false positive rate: If the camera sees a non-terrorist, a bell will not ring 99% of the time, but it will ring 1% of the time

When a resident triggers the alarm, what is the chance that the person is a terrorist? $P(\text{Terrorist}|\text{Ring})$?

3. **Marketing Attribution**

Company Comp Inc. reaches its customers via different promotional channels using targeting. Customer A receives more catalogs than emails and coupons.

In a certain time period, Customer A receives five catalogs, three emails and two coupons; suppose these three formats have similar unit level effect. In other words, the prior probability is proportional to the number of items in each format. Using a separate modeling analysis, we obtain that for customer A, probability of conversion by catalog is 20%, by email is 5%, and by coupons is 10%. When A made a purchase, how do we attribute each purchase? Comment on what *modeling* methodology could help us obtain these conversion probabilities?

Reference

1. The Theory That Would Not Die: How Bayes' Rule Cracked the Enigma Code, Hunted Down Russian Submarines, and Emerged Triumphant from Two Centuries of Controversy Paperback—September 25, 2012. *Sharon Bertsch McGrayne.*

CHAPTER 10

Bayesian Updating and Experimental Design

Learning Objectives

1. Explain the concept of subjective probability and its role in Bayesian reasoning.
2. Comprehend the concept and process of Bayesian updating.
3. Apply Thompson Sampling to the design and analysis of experiments.
4. Describe how Thompson Sampling reduces the opportunity costs in experimental design.

Introduction

The previous chapter introduced Bayes' theorem, its foundation in inverse probability, and its application in addressing the marketing attribution problem. In this chapter, we extend those concepts by exploring Bayesian updating and its application in enhancing the efficiency of field experiments.

Before we get into that, let's first review a fundamental question:

What Is Probability?

To better grasp Bayes' theorem, it is essential first to understand the concept of probability.

In classical statistics, probability is interpreted through the lens of repeated events and long-run frequencies.

Consider a simple example: tossing a regular quarter. What is the probability of the coin landing face up? Intuitively, we know the answer is 50%, or half of

Supplementary Information The online version contains supplementary material available at https://doi.org/10.1007/978-3-032-11130-2_10.

© The Author(s), under exclusive license to Springer Nature Switzerland AG 2026
X. Dong, *Marketing Analytics and Data Science*,
https://doi.org/10.1007/978-3-032-11130-2_10

the times. However, if you toss the coin only twice, can you guarantee that it will land face up exactly once? Certainly not. However, if you repeat this experiment thousands of times, the proportion of face-up outcomes will likely approach 50%.

This idea highlights, in the frequentist sense, how probability is defined as the ratio of:

- Numerator: The number of face-up outcomes
- Denominator: The total number of tosses

The 50% figure becomes meaningful only when the denominator is sufficiently large. When the number of tosses is small, such as only two, the concept of a 50% probability becomes less helpful in describing the observed outcomes. For this reason, this interpretation of probability is often referred to as *frequency-based probability*.

In contrast, Bayesian statistics defines probability as a degree of belief, also known as subjective probability.

For example, a weather forecast predicts a 70% chance of rain tomorrow. Unlike coin tosses, the exact day tomorrow occurs only once, so there is no possibility of repeating the event to compute a frequency. In this case, the probability reflects our belief, informed by available data and expert models, regarding the likelihood of rain.

For much of the twentieth century, particularly from the 1940s to the 1980s, this subjective interpretation of probability faced intense criticism from classical statisticians. The debate between frequentist and Bayesian approaches was once a central topic in discussions within the field. However, in recent decades, this divide has completely disappeared. The growing recognition of Bayesian methods, driven by their effectiveness in addressing complex, real-world problems, has led to widespread acceptance of subjective probability as a valid and powerful tool for decision-making.

Leveraging the concept of subjective probability, the process described in Bayes' theorem can be interpreted as:

- Before collecting data, I have some *prior* belief regarding the answer I'm looking for.
- Then I collect some data and utilize the information in the data to *update* my belief.
- The updated belief is called the *posterior*.

Recall the Bayes theorem equation was written as

$$P(A|B) = \frac{P(B|A) \times P(A)}{P(B)}$$

In this equation,

- $P(A)$ is the prior probability regarding A before collecting any data.
- $P(B|A)$ refers to the data collected that contains information related to A.
- $P(A|B)$ refers to the posterior probability of A given the data collected.

Based on the above understanding, the Bayes theorem describes the relationship among the prior, data, and posterior as

$$\text{Posterior} \propto \text{Prior} \times \text{Collected Data}$$

This is an alternative way of expressing Bayes' theorem. There are two main differences between this equation and the Bayes' theorem equation in the above:

- In this new equation, the right-hand side denominator P(B) contains no information regarding the event A, and is treated as a scaler for the posterior distribution. It is therefore omitted, to simplify the equation.
- Instead of using "=", the new equation uses "∝", standing for "proportional to". It reflects the fact that the right-hand side eliminates the division of a constant and therefore is rescaled.

Bayesian Updating

Returning to the thought experiment described in the previous chapter, Bayes aimed to infer the underlying cause, which is the probability representing where the ball had landed on the board.

To begin, he formulated a prior belief. Without any data, and assuming the chessboard is flat, it is reasonable to believe that the ball has an equal chance of landing anywhere on the board. Thus, the prior probability takes the form of a uniform distribution across all possible landing points on the board.

Next, Bayes began collecting data to update this belief. He did so by asking a colleague to throw a random pebble onto the chessboard and report whether it landed to the right or left of the ball's position. Each pebble toss provided helpful information regarding where the ball landed. For example, after being told twice that the pebble landed to the right of the ball, Bayes could begin to adjust his belief that the ball was more likely to have landed on the left side of the board.

However, with only two observations, considerable uncertainty remained. To reduce this uncertainty, he continued gathering data. Imagine, for instance, that his colleague repeated this process 1000 times and that every single time, the pebble landed to the right of the ball. Such overwhelming evidence would strongly suggest that the ball was positioned near the left edge of the board.

In essence, the more observations Bayes collected, the more confident he became in his posterior inference regarding the cause, which is the ball's actual landing position. This process illustrates a fundamental principle of Bayesian reasoning: as additional data are gathered, beliefs are progressively refined, and uncertainty is reduced.

In this example, the process of collecting data involves obtaining an updated posterior and reducing uncertainty. This differs from the frequentist concept of probabilities, where the value of a probability only makes sense when the sample size is sufficiently large.

This process is also known as "Bayesian Updating" or "Bayesian Learning".

Bayesian updating takes a different approach from the traditional frequentist view. Instead of relying on large amounts of data, it allows us to ***gradually*** improve our understanding as we collect more information. This makes it especially useful when we only have a small amount of data.

One practical example is in experimental design, where methods like Thompson sampling use the concept of Bayesian updating to reduce the cost of experimentation.

Experimental Design and Thompson Sampling

Running large-scale online experiments has become the "Holy Grail" for addressing many challenges. For example, when designing a webpage, where should the "Subscribe" button be placed to maximize the likelihood that visitors will click and subscribe? The most effective way to answer this question is through experimentation. By randomly assigning visitors to different versions of the webpage, each featuring the "Subscribe" button in a different location. Then, comparing the resulting conversion rates, we can identify the most effective design.

Although online platforms make it feasible and less costly to conduct randomized experiments, the *opportunity cost* can still be very high.

For example, if there were two designs, A and B, and each one was randomly assigned 1000 visitors.

- Design A: conversion rate 1%, 10 users converted
- Design B: conversion rate 10%, 100 users converted

Through this experiment, a total of 110 users converted, after 2000 users visited the webpage. Additionally, we found that design B is significantly more effective than design A in facilitating user sign-up.

However, if we hadn't run the experiment and used only design B, we would have gotten 200 (among the 2000) users to sign up. The loss of 90 (200 − 110 = 90) users is due to running the experiment where design A was also included. This loss is characterized as the opportunity cost: we would not have lost these users if we hadn't experimented. However, if we hadn't experimented, we wouldn't have known which design was better in the first place. To resolve this dilemma and to reduce the opportunity cost, we can utilize the Thompson Sampling approach in experimentation.

In Thompson Sampling, we monitor the conversion rate as the experiment is on-going, and we will hopefully learn which design is better before sending the assigned number of visitors to each.

Next, we use an example to illustrate the process and its applications in designing the experiment to reduce the opportunity cost.

An Example

We are designing a webpage that allows visitors to subscribe to a service the website provides. To achieve this, we conduct a random experiment to identify the design with the highest conversion rate.

Suppose we have three designs, named Design A, Design B, and Design C (Fig. 10.1).

In a classical experiment, we would randomly assign a certain number of visitors (say, 100 each). Then wait until each design receives 100 visitors, and compare the number of users that subscribed (say, 30, 10, 5). Design A has a conversion rate of 30/100 = 30% and is the clear winner among the three designs.

Despite the high opportunity cost, it is always important to be able to conduct an online experiment. The question is whether we can design an experiment that reduces opportunity cost as much as possible.

In doing so, we need to balance two opposing forces:

1. To learn about the quality of each design in a reliable way.
2. To only require the least amount of data to learn about the worst design as quickly as possible.

To achieve both goals, we should allocate visitors dynamically based on what we have learned about each design so far. The key idea is to identify worse performing designs early and assign them as few users as possible, thereby reducing the opportunity cost of sending visitors to an inferior option instead of a

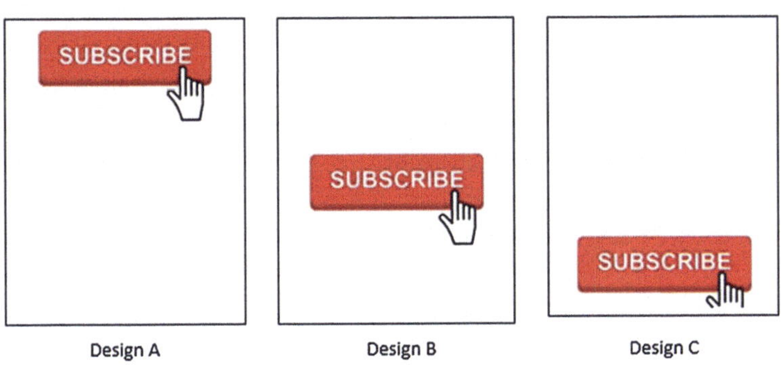

Fig. 10.1 Three designs of the subscribe button

better one. To achieve this, we apply the Thompson Sampling method, which leverages the concept of Bayesian updating.

Thompson Sampling

Thompson sampling leverages Bayesian updating: as new outcomes arrive, we update the posterior distribution of each design's conversion rate and then randomly select the next design in proportion to how likely it is to be the best given the current evidence. This approach naturally balances exploration (testing uncertain options) with exploitation (favoring designs that already look promising). The Bayesian Updating method allows us to check the conversion rate while running the experiment, rather than waiting for the whole experiment to finish to check the conversion rate.

Just like any Bayesian method, to apply the Thompson Sampling approach, we need to begin with a prior distribution for the unknown quantity. The unknown parameters here are the conversion rates for each design, denoted as θ_A, θ_B, and θ_C.

Given that we have no prior information about each design's conversion probability, other than that they must lie between 0 and 1, we assume each of them has a prior distribution that is uniformly distributed between 0 and 1.

The prior distributions before the experiment starts:

- $\theta_A \sim U(0, 1)$, with a probability distribution function (PDF) $f(\theta_A) = 1$
- $\theta_B \sim U(0, 1)$, with PDF $f(\theta_B) = 1$
- $\theta_C \sim U(0, 1)$, with PDF $f(\theta_C) = 1$

In other words, before running the experiment, we have no clue which design is better. All of them are equally possible to be in the range from 0 to 1.

Now, we start collecting data through the experiment. The key is that, with Bayesian updating, we do not need to wait for all the data to be collected at once; we can start with a small number of data points and then update the distribution for the conversion rates. Suppose we first assign each design to only 10 visitors, and get the following data in the first round:

The data from the first round:

- Design A: 10 visitors, 4 converted
- Design B: 10 visitors, 3 converted
- Design C: 10 visitors, 2 converted

These data provide information on what we have learned about the conversion rate so far for each design. The next question is how to incorporate that evidence into Bayes' rule, in other words, how to represent the "collected data" item in the Bayesian updating equation. The answer is using the likelihood function: for each design, the likelihood is the joint probability of the observed outcomes so far, given the value of a conversion rate.

The likelihood functions for each design in Round 1 can be expressed as $L(\text{Data}_1|\theta)$.

- Design A: the likelihood function is $L_{A1} = \theta_A^4 \times (1-\theta_A)^6$
- Design B: the likelihood function is $L_{B1} = \theta_B^3 \times (1-\theta_B)^7$
- Design C: the likelihood function is $L_{C1} = \theta_C^2 \times (1-\theta_C)^8$

This formulation is also called the **Beta distribution, which is explained next.**

Beta Distribution

Beta distribution is often used to model uncertainty about a probability metric, such as the conversion rate in this example. It represents the scenario where we conducted N independent trials with only two possible outcomes: success and failure. If we observe p successes (conversions in our example) and $(N - p)$ failures (no conversions), the underlying conversion probability can be described with a Beta distribution, which captures not only a point estimate like. p/N, but also the uncertainty around the estimate given the amount of the data observed so far.

The Beta distribution is defined on the support between 0 and 1, the same range as a probability value can take. It has two parameters α and β. It can be represented as Beta(α,β).

The distribution function of Beta(α,β) is written as

$$\text{PDF}(x) \propto x^{(\alpha-1)}(1-x)^{(\beta-1)}$$

In this notation, the distribution can be interpreted as:

- x represents the success rate, for which we are interested. In our example, it is the conversion rate for each design that we are trying to figure out with the experiment
- the number of successful trials is $p = \alpha - 1$,
- the number of failure trials is $\beta - 1$,
- the total number of trials $N = \alpha + \beta - 2$.

Based on these two parameters α and β, we can calculate the mean and standard deviation of the beta distribution:

$$\text{Mean} = \frac{\alpha}{\alpha+\beta}$$

This is one of those distributions where the mode of the distribution (the point at which the distribution reaches its peak) is different from its mean. The mode of the distribution can be calculated as

$$\text{Mode} = \frac{\alpha - 1}{\alpha + \beta - 2}, \alpha > 1, \beta > 1$$

Note that, based on the above description and parameter definition, we know that $p = \alpha - 1$ and $N = \alpha + \beta - 2$, so the mode is actually calculated as the ratio of p/N.

When $\alpha = 1$ and $\beta = 1$, the mode value cannot be calculated. In fact, the distribution becomes a uniform distribution.

When $\alpha < 1$ or $\beta < 1$, it has no mode.

Its variance can be calculated as

$$\text{Variance} = \frac{\alpha\beta}{(\alpha + \beta)^2 (\alpha + \beta + 1)}$$

Note that the variance is inversely related to the value of $(\alpha + \beta)$. As $(\alpha + \beta)$ increases, the variance decreases. Since $(\alpha + \beta) - 2 = N$, that is $(\alpha + \beta) = N + 2$, this relationship indicates that a larger number of trials N leads to a smaller variance and, therefore, a more certain distribution. This is consistent with the intuitive understanding that collecting more data enhances our confidence in the underlying distribution.

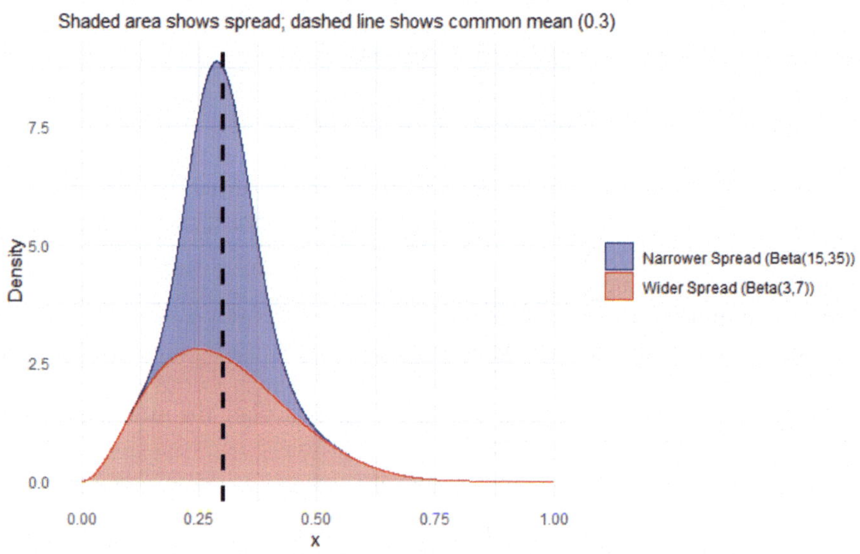

Fig. 10.2 Illustration of Beta distribution with same mean but different spread

As an illustrative example, Fig. 10.2 displays two Beta distributions: Beta(3,7) and Beta(15,35). Both have the same mean value of 0.3 (indicated by the black dashed line), but differ in other important ways. Notably, they have slightly different modes and substantially different standard deviations.

The value $\alpha + \beta - 2$ can be interpreted as reflecting the number of data points or trials underlying each distribution. The red-shaded area corresponds to the Beta(3,7) distribution, which reflects fewer data points and thus exhibits a wider spread. This broader and flatter shape of the distribution implies that although the mean is 0.3, the probability of observing values near 0.1 or 0.3 does not differ dramatically.

In contrast, the blue-shaded area corresponds to the Beta(15,35) distribution, which is based on a much larger number of trials. As a result, it is far more concentrated around the mean, and the likelihood of observing other values rather than the mean 0.3, such as 0.1, is much lower. This demonstrates how larger data samples lead to tighter distributions and greater certainty about the underlying probability.

Now, back to our example for the webpage design experiment:

The posterior distributions for these three conversion rates can be obtained by

$$\text{Prior} \times \text{Likelihood}$$

Design A: the posterior distribution of θ_A after the first round is

$$\begin{aligned} P(\theta_A | \text{Round 1}) &\propto \text{Prior}_{A1} \times L_A^1 \\ &= 1 \times \theta_A^4 \times (1-\theta_A)^6 \\ &= \theta_A^4 \times (1-\theta_A)^6 \end{aligned}$$

Design B: the posterior distribution of θ_B after the first round is

$$\begin{aligned} P(\theta_B | \text{Round 1}) &\propto \text{Prior}_{B1} \times L_B^1 \\ &= 1 \times \theta_B^3 \times (1-\theta_B)^7 \\ &= \theta_B^3 \times (1-\theta_B)^7 \end{aligned}$$

Design C: the posterior distribution of θ_C after the first round is

$$\begin{aligned} P(\theta_C | \text{Round 1}) &\propto \text{Prior}_{C1} \times L_C^1 \\ &= 1 \times \theta_C^2 \times (1-\theta_C)^8 \\ &= \theta_C^2 \times (1-\theta_C)^8 \end{aligned}$$

Note that the powers are $\alpha - 1$ and $\beta - 1$ in the PDF. Therefore,

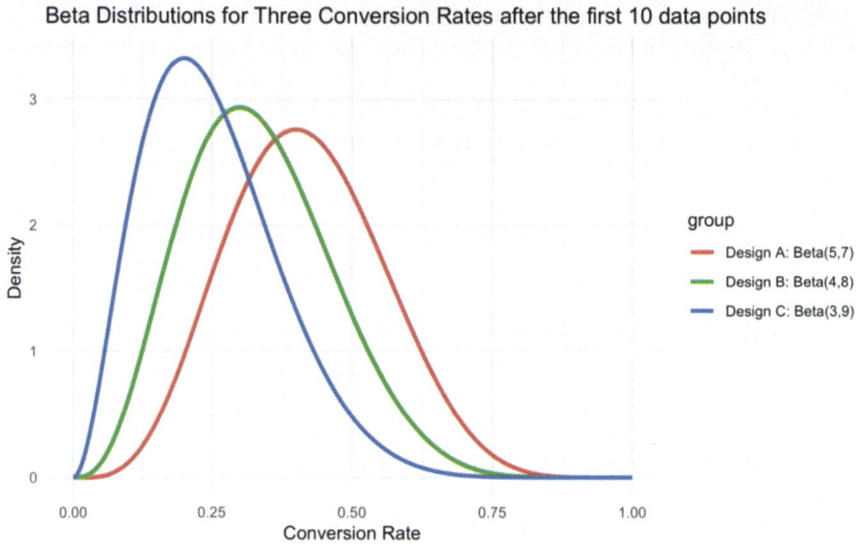

Fig. 10.3 Distribution of the conversion rates after the first round

- The posterior distribution for Design A, after 10 data points, is Beta(5, 7).
- The posterior distribution for Design B, after 10 data points, is Beta(4, 8).
- The posterior distribution for Design C, after 10 data points, is Beta(3, 11).

These three distributions are plotted below (Fig. 10.3):

As shown in the plot, the three distributions largely overlap, making it difficult to confidently determine which design performs the best. While Design A appears to be slightly better than B, which in turn seems to be better than C, the differences are subtle and uncertain. With only 10 data points collected for each design, the level of uncertainty remains high, as indicated by the significant overlap among the distributions. Therefore, additional data collection is necessary before drawing any firm conclusions.

Despite the considerable uncertainty, even this small dataset provides valuable insights into the relative performance of each design. These insights can inform our data allocation strategy for the next round. Specifically, we can assign more users to the designs that are more likely to perform well while allocating fewer users to those that appear less promising.

To achieve this, we can use the posterior distributions of the three conversion rates to calculate *the probability that each design is the best.* Then we can allocate the number of users *based on the likelihood of each design being the best.*

To obtain that probability, we have to rely on simulations.

```
# Set seed for reproducibility
set.seed(123)

# Define parameters
alpha <- c(5, 4, 3)
beta <- c(7, 8, 9)

# Number of samples
n_samples <- 100000

# Sample from the beta distributions
samples_1 <- rbeta(n_samples, alpha[1], beta[1])
samples_2 <- rbeta(n_samples, alpha[2], beta[2])
samples_3 <- rbeta(n_samples, alpha[3], beta[3])

# Combine samples into a matrix
samples <- cbind(samples_1, samples_2, samples_3)

# Find which sample is the maximum in each draw
best_design <- apply(samples, 1, which.max)

# Calculate probabilities
prob_1 <- mean(best_design == 1)
prob_2 <- mean(best_design == 2)
prob_3 <- mean(best_design == 3)

# Print the results
cat("Probability Design 1 is best:", prob_1, "\n")

## Probability Design 1 is best: 0.60038

cat("Probability Design 2 is best:", prob_2, "\n")

## Probability Design 2 is best: 0.28761

cat("Probability Design 3 is best:", prob_3, "\n")

## Probability Design 3 is best: 0.11201
```

Based on these results, in the next round, when we allocate another 30 users, we allocate them based on each design's probability of being the best, that is:

In Round 2, the number of visitors randomly assigned to each design is

- Design A: $30 \times P_1(\text{ABest}) = 30 \times 0.60038 = 18.0$
- Design B: $30 \times P_1(\text{BBest}) = 30 \times 0.28761 = 8.6$
- Design C: $30 \times P_1(\text{CBest}) = 30 \times 0.11201 = 3.3$

To get integers from the above calculation results, we decided to choose the number of users for the second round for each design as:

- Design A: 18 users
- Design B: 8 users
- Design C: 4 users

Now repeat the above process:

The prior distributions of the conversion rate for Round 2 are the posterior distributions from Round 1.

That is, the prior distributions before Round 2 are:

- Design A: $P(\theta_A|\text{Round1}) = \theta_A^4 \times (1-\theta_A)^6$
- Design B: $P(\theta_B|\text{Round1}) = \theta_B^3 \times (1-\theta_B)^7$
- Design C: $P(\theta_C|\text{Round1}) = \theta_C^2 \times (1-\theta_C)^8$

Then we get the conversion data from Round 2 as follows:

- Design A: 10 converted out of 18 visitors
- Design B: 3 converted out of 8 visitors
- Design C: 1 converted out of 4 visitors

The likelihood functions for each design in Round 2 $L(\text{Data}_2|\theta)$:

- Design A: the likelihood function is $L_{A2} = \theta_A^{10} \times (1-\theta_A)^8$
- Design B: the likelihood function is $L_{B2} = \theta_B^3 \times (1-\theta_B)^5$
- Design C: the likelihood function is $L_{C2} = \theta_C^1 \times (1-\theta_C)^3$

Combining the prior for Round 2, which is also the posterior for Round 1, and the likelihood functions from Round 2, we can obtain the posterior distributions for these three conversion rates after Round 2:

1. Design A: the posterior distribution of θ_A after Round 2 is

$$\begin{aligned} P(\theta_A|\text{Round2}) &= \text{Prior}_{A2} \times L_{A2} \\ &= \theta_A^4 \times (1-\theta_A)^6 \times \theta_A^{10} \times (1-\theta_A)^8 \\ &= \theta_A^{14} \times (1-\theta_A)^{14} \end{aligned}$$

2. Design B: the posterior distribution of θ_B after Round 2 is

$$\begin{aligned} P(\theta_B|\text{Round2}) &= \text{Prior}_{B2} \times L_{B2} \\ &= \theta_B^3 \times (1-\theta_B)^7 \times \theta_B^3 \times (1-\theta_B)^5 \\ &= \theta_B^6 \times (1-\theta_B)^{12} \end{aligned}$$

3. Design C: the posterior distribution of θ_C after Round 2 is

$$\begin{aligned} P(\theta_C | \text{Round } 2) &= \text{Prior}_{C2} \times L_{C2} \\ &= \theta_C^2 \times (1-\theta_C)^8 \times \theta_C^1 \times (1-\theta_C)^3 \\ &= \theta_C^3 \times (1-\theta_C)^{11} \end{aligned}$$

The three posterior distributions after the second round are plotted below (Fig. 10.4).

After this round, we have accumulated a total of 60 data points across the three designs. As shown in the plot, the leading design has become more apparent compared to the previous round, though the result is still not definitive. Therefore, we need to continue with a third round of data collection. Before proceeding, we will calculate the probability of each design being the best based on the current data and use these probabilities to guide the allocation of the following 30 data points.

As the experiment progresses and more data is gathered, we gradually gain greater confidence in identifying the most effective design. At the same time, we aim to minimize opportunity costs by allocating fewer users to the designs that are less likely to perform best.

You will notice that soon, Design C will receive very few visitors. This approach strikes a balance between efficiency and exploration by minimizing wasted opportunities on less promising designs while still allowing each design to demonstrate its potential.

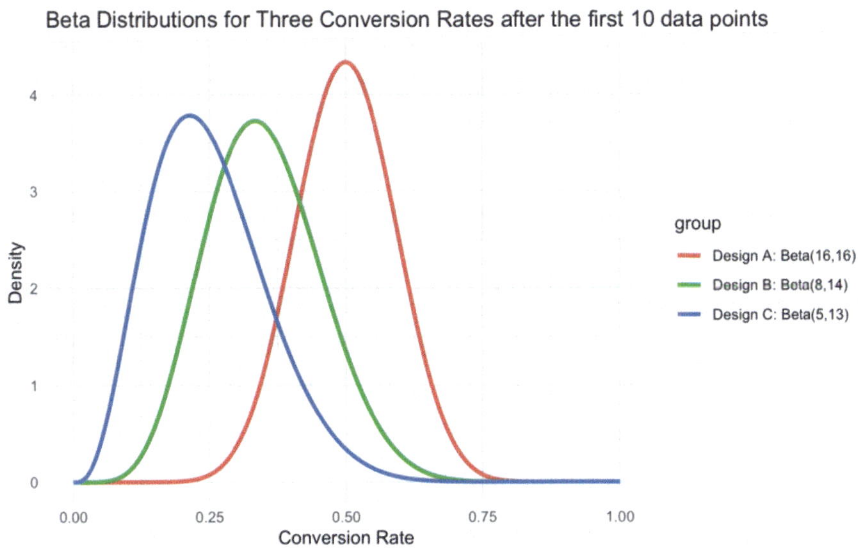

Fig. 10.4 Posterior distributions after Round 2

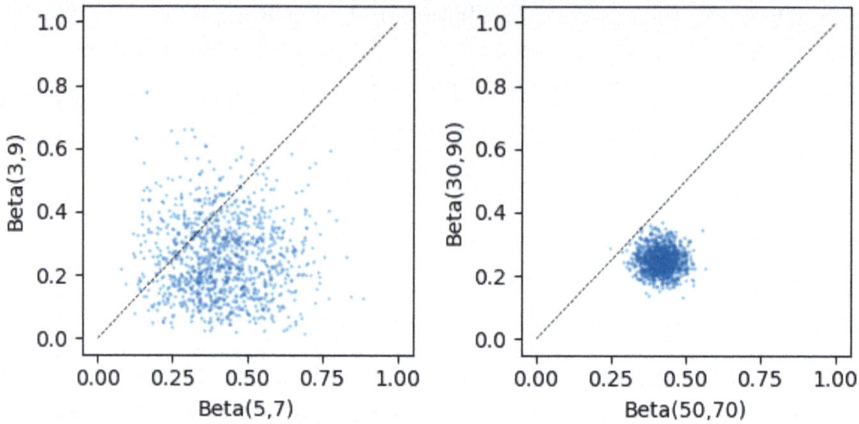

Fig. 10.5 Compare Beta distributions using simulations

Figure 10.5 illustrates how our confidence in identifying the higher conversion rate improves as we collect more data.

On the left, we see a scatter plot comparing two Beta distributions: Beta(5, 7) on the *x*-axis and Beta(3, 9) on the *y*-axis. These parameters reflect a small number of trials, resulting in both distributions being relatively dispersed. As a result, the scatter points are widely spread, with many falling both above and below the 45-degree line, indicating substantial uncertainty about which conversion rate is higher.

On the right, the scatter plot shows simulated points from two Beta distributions with much larger sample sizes: Beta(50, 70) on the *x*-axis and Beta(30, 90) on the *y*-axis (i.e., the parameters from the left plot scaled up by a factor of 10). With more data, the distributions become more concentrated, and the scatter plot becomes significantly tighter. Almost all the points fall below the 45-degree line, clearly indicating that the conversion rate represented on the *x*-axis is very likely higher than the one on the *y*-axis.

This comparison illustrates a key principle: as the size of the collected data increases, uncertainty decreases, enabling us to make more confident probabilistic statements about relative performance. The process of Thompson sampling follows this statistical pattern.

As the opportunity cost could be substantial, companies like Google and Facebook run thousands of experiments on their platforms. The Bayesian updating concept can help reduce the opportunity cost dramatically by using Thompson Sampling.

Summary

These two chapters demonstrate two of the essential features of Bayesian statistics:

- Inverse probability
- Bayesian updating

The first one finds its applications in dealing with one of the most popular marketing analytics problems: marketing attribution. It helps to convert the problem into an inverse probability problem and apply Bayes' theorem directly.

The second one finds its applications in dynamically designing a field experiment aiming at reducing opportunity cost. Such a cost has been one of the biggest obstacles to marketing researchers when evaluating possible alternatives in a controlled experiment.

Bayesian statistics, as an alternative approach to the classical approach, can tackle some of the common and challenging marketing problems from a different angle. In the near future, we will see its even wider applications and contributions to more data-driven marketing decision-making.

Homework

1. Explain how Thompson Sampling reduces opportunity cost.
2. Discuss some other scenarios you may find helpful to apply Bayesian updating or Thompson Sampling.

CHAPTER 11

Causal Analysis with Randomized Field Experiment

Learning Objectives

1. Describe the importance of causal analysis
2. Conduct randomized field experiment using randomized controlled trial method
3. Evaluate the average treatment effect using statistical analysis
4. Estimate heterogeneous treatment effects using regression analysis

Introduction

The study of cause-and-effect relationships dates back to ancient philosophy when early thinkers sought to explain the forces shaping the world around them. In the modern era, the ability to identify causes behind observed phenomena has become fundamental across disciplines such as business, science, and economics. The goals of finding answers to these questions have shifted from satisfying curiosity in ancient times to seeking ways to change the outcomes in modern days.

At the heart of causal analysis lie two central questions:

1. What will happen if we change a particular input?
2. Given an observed outcome, what caused it?

Supplementary Information The online version contains supplementary material available at https://doi.org/10.1007/978-3-032-11130-2_11.

© The Author(s), under exclusive license to Springer Nature Switzerland AG 2026
X. Dong, *Marketing Analytics and Data Science*, https://doi.org/10.1007/978-3-032-11130-2_11

In both cases, the goal is to isolate the incremental effect of a specific action or factor on the outcome. While this can be straightforward in controlled settings, it often proves challenging in more complex scenarios.

For instance, in a cake recipe, if sugar is the only sweet ingredient and the cake turns out overly sweet, the cause is clear: too much sugar. The method to improve is also easy: to reduce the sugar amount next time. However, if both sugar and maple syrup contribute to the sweetness, the situation becomes ambiguous. Should we reduce one, the other, or both? To answer this, we could conduct further tests, adjusting sugar while keeping maple syrup constant and vice versa, to pinpoint the individual impact of each ingredient.

Randomized Field Experiment

The above example highlights two essential conditions for identifying treatment effects:

- First, the causal **factor** must be clearly defined and controllable. This is important because one of the main purposes of causal analysis is to guide action: once we know a factor causes an outcome, we can adjust that factor to achieve a desired result. If the factor cannot be manipulated in practice, then identifying its causal effect offers limited practical value.
- Second, we must be able to compare **outcomes** with and without that factor. In other words, the impact of each factor can be identified through the changes in outcomes.

When feasible, controlled experiments, such as randomized controlled trials (RCTs), are the most popular way to identify causal effects. In an RCT, researchers randomly assign units (such as consumers, stores, or regions) to receive a treatment (e.g., a new ad campaign, a price promotion, or a redesigned app feature) or to a control condition that does not receive it, and then compare outcomes between groups.

Any RCT design needs to satisfy the above two minimal requirements. In practice, the outcomes we are interested in measuring are often influenced by a selection of variables. When designing an experiment, we need to ensure the causal variable can be directly and feasibly changed to observe its effect. For example, in the context of digital marketing, whether a message was sent to the user or not is controllable, however, the number of users visiting a website each day is not directly changeable.

When analyzing RCT outcomes, we are generally interested in evaluating two types of treatment effects:

- The average treatment effect (ATE), which evaluates the average outcome differences across the treatment and control groups.
- The heterogeneous treatment effect (HTE), which evaluates differential treatment effects across different segments. The segments are usually pre-

defined based on some observed characteristics of the experiment participants.

Running Randomized Field Experiment Using RCT

A field experiment is an RCT conducted in a real-world setting, like a live marketplace or an operating business, where participants behave naturally and the intervention is implemented under practical constraints; as a result, field experiments often provide strong causal evidence with high managerial relevance, while still relying on randomization to support credible causal inference.

When conducting a randomized field experiment, participants in the treatment and control groups should be statistically equivalent. This does not imply that each individual must have an identical counterpart in the other group; rather, the two groups should be similar on average in both observable and unobservable characteristics.

To achieve this statistical similarity, researchers use random assignment. As the name suggests, an RCT randomly assigns each participant to either the treatment or the control group. This means every participant has an equal chance of being placed in either group. Random assignment helps satisfy two critical conditions:

1. The groups are statistically equivalent at the start of the experiment.
2. Any differences observed after the intervention can be attributed to the treatment, not to pre-existing differences between the groups.

Two Key Conditions for Valid Random Assignment

1. Avoid systematic bias in assignment

   Random assignment must be free from any patterns that could correlate with the outcome. For example, when conducting an experiment on a website, assigning by user IDs can be risky if these IDs reflect signup order: earlier users (with lower IDs) may be more loyal. Suppose one group consists mainly of early users and the other mainly of late users. In that case, the groups differ in loyalty before the experiment begins, which will bias the results, especially when loyalty affects the outcome.
2. Ensure sufficient sample size

   Randomness works best when each group has many participants. If a group has only a few person, even random assignment can lead to big differences. With larger groups, individual differences tend to cancel each other out, making the groups more likely to be statistically similar.

Once random assignment is complete, one group is designated as the control group (which receives no intervention), and the other one as the treatment group (which receives the intervention). It is also possible to have multiple treatment groups, each receiving a different intervention. The goal is to compare outcomes between these groups to determine whether the treatment caused any changes.

One possible undesirable outcome is that the treatment effects can be contaminated by the "placebo effect". This concept is originated in the medical field. It speaks to the possible scenario, where a patient believe they are receiving an effective treatment, even if the treatment itself has no active ingredient (or no direct causal mechanism). This happens because expectations can influence perception, attention, and other psychological and physiological responses. For example, someone may report less pain after taking a "pill" that is actually a sugar tablet. In experiments, placebo effects matter because improvements in the treatment group might reflect belief and context rather than the treatment's true impact. Therefore, it is often important to use placebo controls by giving the control group an experience that looks and feels similar to the treatment, just without the active "ingredient." For example, if the treatment is mailing a coupon, the placebo control could be sending a similar mailer, only without the coupon. In other words, the "control group" does not always imply that we do nothing with them, in a randomized field experiment.

Before comparing the statistics between these outcomes, it's important to perform a randomness check between the two groups by comparing some baseline statistics to ensure they are not statistically different. This involves checking for differences in key observables, such as age, prior behavior, or other characteristics. If the groups are statistically similar before running the experiment, then any differences observed after the treatment can be attributed to the treatment itself.

An Example

In this example, we examine an online learning platform where users can sign up to access resources that help them prepare for a certification exam. Those who pass the exam receive a nationally recognized certificate[1] which can enhance their job prospects. The platform offers on-demand access to video lectures and practice tests, allowing users to study at their own pace.

One challenge the platform observed was a low level of engagement from a large number of users, as measured by their average hours spent studying on this platform each week. To address this issue and to encourage continued engagement, the platform introduced a social comparison feature. Upon logging in, users now see a message alongside their personal study time that reads, "On average, people in your class have studied 5 hours this week". The goal of this intervention is to subtly motivate users by showing how their study habits compare to their peers. This experiment was conducted over two months. To evaluate its influence on user engagement, we use a sample dataset collected from this controlled experiment.

[1] Due to a confidentiality agreement with the platform, the identity of the company cannot be disclosed.

```
causal_df <- read.csv('Causal_SocialComparison.csv')
library(dplyr)

causal_df %>% count(Group)

##       Group  n
## 1   Control 70
## 2 Treatment 70
```

In this dataset, a random sample of 70 users from the control group and another 70 from the treatment group is selected.

```
head(causal_df)

##   UserID     Group weekly_hour Pre_hour Age
## 1  38908 Treatment      4.6907    50.52  22
## 2  39080 Treatment      4.4387    49.59  28
## 3  39206   Control      2.8347    36.10  28
## 4  39290 Treatment      4.5796    45.33  22
## 5  39370   Control      3.4767    40.63  39
## 6  39408 Treatment      4.2172    38.93  22
```

This dataset contains five columns:

- The first column identifies each user in each row.
- The second column indicates group assignment, with users placed either in the control group (received the same message as before) or in the treatment group (received an additional message highlighting the average study time of their peers).

The remaining three columns fall into two categories:

- The outcome variable: weekly_hour, which measures each user's average number of hours spent studying per week.
- User characteristics: These variables indicate the basic demographic information of each user and their interactions with the learning platforms before the intervention. These variables can be used for the randomness check, that helps us assess whether there are any systematic differences between the treatment and control groups based on these measurable observed factors.

To compare the two groups, we evaluate the basic statistics, mean, and standard deviations, for each variable.

```
summary_stats <- causal_df %>%
  group_by(Group) %>%
  summarise(
    number_users    = n(),
    weekly_hour_mn  = mean(weekly_hour, na.rm = TRUE),
    weekly_hour_sd  = sd(weekly_hour, na.rm = TRUE),
    Age_mn          = mean(Age, na.rm = TRUE),
    Age_sd          = sd(Age, na.rm = TRUE),
    Pre_hour_mn     = mean(Pre_hour, na.rm = TRUE),
    Pre_hour_sd     = sd(Pre_hour, na.rm = TRUE)
  )

# View the result
print(summary_stats)

## # A tibble: 2 × 8
##   Group     number_users weekly_hour_mn weekly_hour_sd Age_mn Age_sd
Pre_hour_mn
##   <chr>            <int>          <dbl>          <dbl>  <dbl>  <dbl>
<dbl>
## 1 Control             70           3.49          0.346   25.2   4.20
38.5
## 2 Treatment           70           4.56          0.370   26.2   5.47
41.7
## # ℹ 1 more variable: Pre_hour_sd <dbl>
```

To visualize the differences between the two groups, we can create **kernel density estimate** (KDE) plots for each variable.

KDE plots can be viewed as a smoothed alternative to traditional histograms. Unlike histograms, which require manually selecting bin widths and intervals, KDE plots avoid this arbitrary binning by using a non-parametric approach. They do not assume any specific underlying distribution; instead, they allow the observed data to shape the estimated distribution. The result is a smooth curve that closely resembles a probability density function (PDF), offering a more refined view of how the data is distributed.

These plots illustrate how the distributions of the treatment and control groups compare and the extent to which they overlap.

```
library(ggplot2)

## Warning: package 'ggplot2' was built under R version 4.4.3

# Assuming your data frame is named df and group_number is a factor
causal_df$Group <- as.factor(causal_df$Group)

# Kernel density plot for weekly_hour
ggplot(causal_df, aes(x = weekly_hour, fill = Group, color = Group)) +
  geom_density(alpha = 0.4) +
  labs(title = "Kernel Density Plot: Weekly Hours",
       x = "Weekly Hours",
       y = "Density",
       fill = "Group",
       color = "Group") +
  theme_minimal()


# Kernel density plot for Age
ggplot(causal_df, aes(x = Age, fill = Group, color = Group)) +
  geom_density(alpha = 0.4) +
  labs(title = "Kernel Density Plot: User Age",
       x = "User Age",
       y = "Density",
       fill = "Group",
       color = "Group") +
  theme_minimal()


# Kernel density plot for Pre_hour
ggplot(causal_df, aes(x = Pre_hour, fill = Group, color = Group)) +
  geom_density(alpha = 0.4) +
  labs(title = "Kernel Density Plot: Pre-experiment Study Hours",
       x = "Pre-experiment Study Hours",
       y = "Density",
       fill = "Group",
       color = "Group") +
  theme_minimal()
```

```
weekly_hour_ttest <- t.test(weekly_hour ~ Group, data = causal_df)
print(weekly_hour_ttest)

##
##  Welch Two Sample t-test
##
## data:  weekly_hour by Group
## t = -17.822, df = 137.37, p-value < 2.2e-16
## alternative hypothesis: true difference in means between group Control and
group Treatment is not equal to 0
## 95 percent confidence interval:
##  -1.1975662 -0.9583652
## sample estimates:
##   mean in group Control mean in group Treatment
##                3.486636                 4.564601
```

Based on the plot shown in Figure 11.1, it is evident that the two groups differ substantially in their study hours per week. Based on the plots in Figure 11.2 and Figure 11.3, the treatment and control groups but appear quite similar across the other two variables, including age and study hours measured before conducting the experiment.

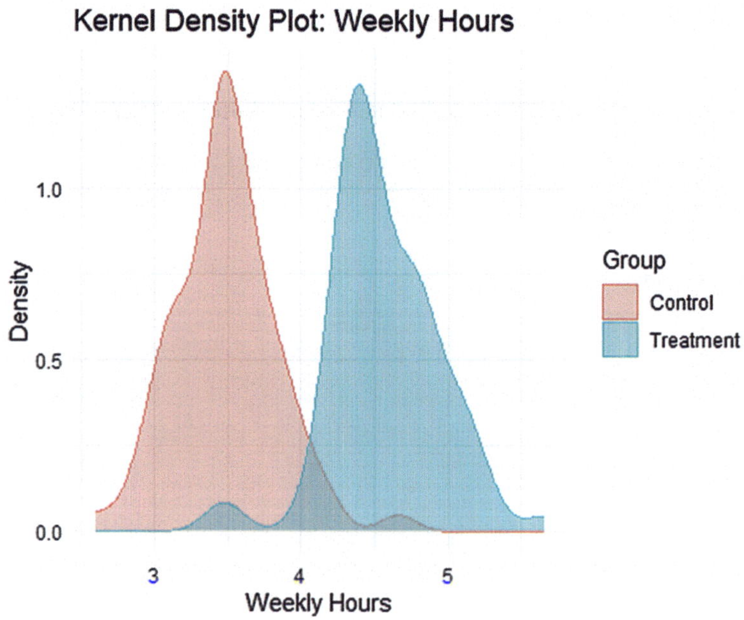

Fig. 11.1

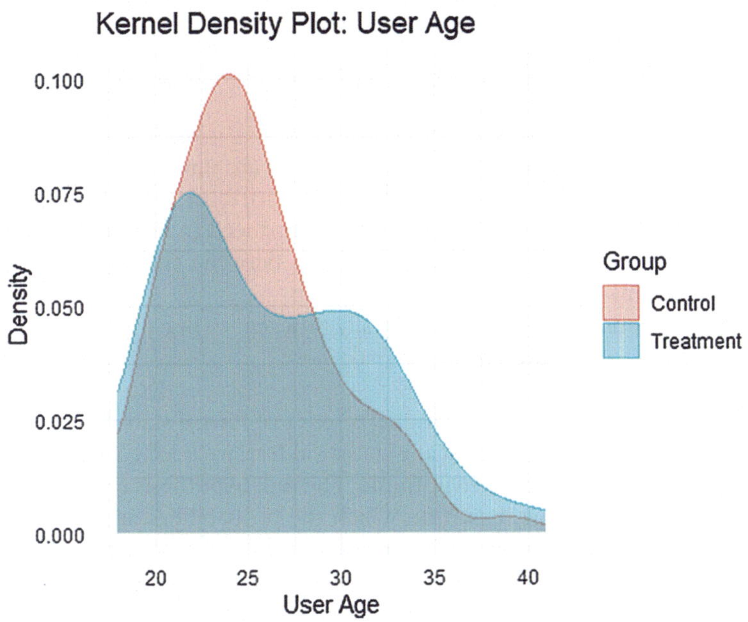

Fig. 11.2

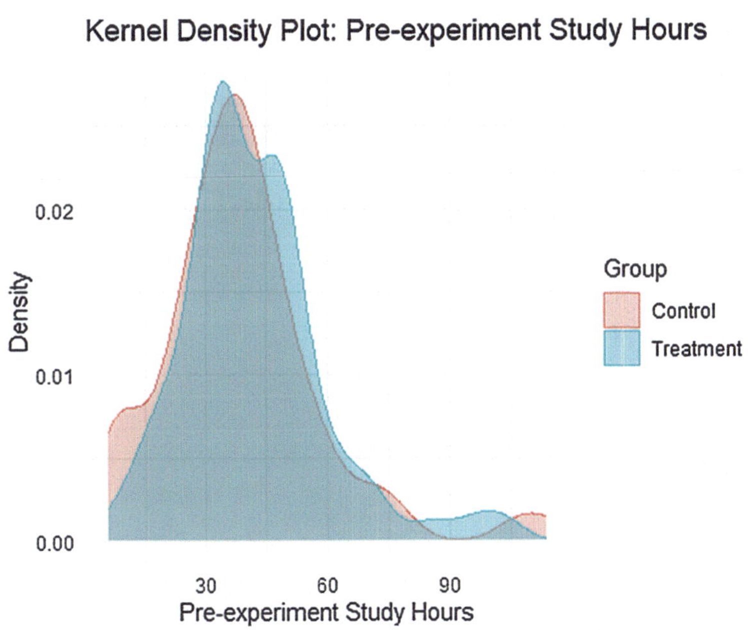

Fig. 11.3

Formal Statistical Test

In any randomized field experiment, we aim to achieve two goals at the same time:

- The control group and the treatment group are statistically similar before the experiment. This ensures that we can attribute the difference in the outcome variable to the intervention introduced in the experiment, i.e., finding the cause-and-effect relationship between the intervention and the outcome variable.
- The outcome variables are statistically different to illustrate that the intervention makes a difference.

To achieve both, we need to conduct a series of formal statistical tests, **two-sample t-test**. Before conducting any statistical test, we need to specify the null hypothesis and the alternative hypothesis. The null hypothesis refers to the case that we assume to be true, based on which the probabilities are calculated. In this example, the null hypothesis is that there is **no** difference between the two groups. Under this assumption, we can calculate the probability of observing the statistics in the data; the probability is the p-value.

To understand the idea of the null hypothesis, it can be viewed as similar to the condition in the conditional probability discussed in Chaps. 9 and 10.

To compare weekly_hour across the two groups, we calculate the average values separately for the control and treatment groups. Although the means differ slightly, a two-sample t-test is necessary to determine whether that difference is statistically significant. The plot in Figure 11.1 suggests limited overlap, but the t-test quantifies the likelihood that such a difference could arise by chance.

This can be achieved using the following R code:

```
Age_ttest <- t.test(Age ~ Group, data = causal_df)
print(Age_ttest)

##
##  Welch Two Sample t-test
##
## data:  Age by Group
## t = -1.1443, df = 129.44, p-value = 0.2546
## alternative hypothesis: true difference in means between group Control and
group Treatment is not equal to 0
## 95 percent confidence interval:
##  -2.5730107  0.6872965
## sample estimates:
##   mean in group Control mean in group Treatment
##              25.21429                26.15714

Pre_hour_ttest <- t.test(Pre_hour ~ Group, data = causal_df)
print(Pre_hour_ttest)

##
##  Welch Two Sample t-test
##
## data:  Pre_hour by Group
## t = -0.98647, df = 136.02, p-value = 0.3257
## alternative hypothesis: true difference in means between group Control and
group Treatment is not equal to 0
## 95 percent confidence interval:
##  -9.444155  3.157869
## sample estimates:
##   mean in group Control mean in group Treatment
##              38.54586                41.68900
```

One important metric among these results is the *p*-value. It indicates that if the two samples are the same, the probability that we would obtain the current data sample.

- If the *p*-value is very small (typically less than 0.05), it means we are pretty certain that these two samples (control vs. treatment groups) are statistically different.
- If the *p*-value is quite large, it means we are not certain that these two samples are statistically different.

As shown above, the *p*-value for the weekly_hour variable across the two groups is quite small, so we reject the null hypothesis of equal means and conclude that the treatment and control groups are statistically different. This indicates that the intervention produced a statistically significant increase in study time on the platform.

To validate the experiment, we also verify baseline equivalence: using similar t-tests on the two observed pre-treatment variables, we check that the groups do not differ significantly in the absence of the intervention.

> The results above show that, for the average age, despite the values are different between these two groups, 25.2 for the control group and 26.2 for the treatment group, the *p*-value from the t-test is 0.25, much bigger than 0.05, we therefore conclude that these two groups are not statistically different in their age.

Similar results are obtained for the pre-experiment study hour; despite the difference in the mean of `pre_hour`, the *p*-value is close to 0.33. We conclude that these two groups are not statistically different in the pre-hour.

Based on the above three t-tests, we conclude that the two groups, control and treatment, are statistically similar, and the intervention drives a statistically significant increase in average study hours per week.

Heterogeneous Treatment Effects: Using a Regression Model

The analysis above using the t-test provides an estimate of the average treatment effect (ATE), showing whether the treatment had a statistically significant impact on average. However, to understand how different users may respond differently to the treatment, which is also known as heterogeneous treatment effects (HTE), we need a more flexible approach, such as a regression model.

In the regression model, we include a dummy variable that indicates whether a user belongs to the treatment group or the control group. It takes value 1 if in the treatment group and 0 if in the control group. The coefficient on this variable represents the average difference in outcomes between the two groups, which is also the ATE.

One observation from our dataset is that users span a wide age range. We hypothesize that users of different ages may react differently to social comparison messages. To test this, we create another dummy variable: it equals to 1 if the user is older than 30, and 0 otherwise.

We then include this age dummy and an interaction term between the age dummy and the treatment dummy in the regression model. The coefficient on the interaction term captures how the treatment effect differs for users over 30 compared to younger users.

```
causal_df$Treat_dm <- ifelse(causal_df$Group == "Treatment", 1, 0)
causal_df$Age_gt30 <- ifelse(causal_df$Age>30, 1, 0)
regr <- lm(weekly_hour ~ Treat_dm + Age + Pre_hour + Treat_dm*Age_gt30, data
= causal_df)
summary(regr)

##
## Call:
## lm(formula = weekly_hour ~ Treat_dm + Age + Pre_hour + Treat_dm *
##     Age_gt30, data = causal_df)
##
## Residuals:
##      Min       1Q   Median       3Q      Max
## -0.86844 -0.23800 -0.00657  0.22669  1.08402
##
## Coefficients:
##                    Estimate Std. Error t value Pr(>|t|)
## (Intercept)        3.708413   0.244198  15.186  < 2e-16 ***
## Treat_dm           1.142915   0.065958  17.328  < 2e-16 ***
## Age               -0.015285   0.009815  -1.557  0.12176
## Pre_hour           0.002852   0.001567   1.821  0.07090 .
## Age_gt30           0.375817   0.148085   2.538  0.01230 *
## Treat_dm:Age_gt30 -0.398429   0.151911  -2.623  0.00973 **
## ---
## Signif. codes:  0 '***' 0.001 '**' 0.01 '*' 0.05 '.' 0.1 ' ' 1
##
## Residual standard error: 0.3462 on 134 degrees of freedom
## Multiple R-squared:  0.7247, Adjusted R-squared:  0.7145
## F-statistic: 70.56 on 5 and 134 DF,  p-value: < 2.2e-16
```

The interpretations of the parameters from this model:

- The parameter for Treat_dm: measures the estimated average treatment effect. As indicated by its *p*-value, it is statistically significant and positive.
- The parameter for age: this is not statistically significant, indicating that a user's age is not directly correlated with the time they study each week.
- The parameter for Pre_hour: the *p*-value is 0.07, indicating a very weak impact from this variable.
- The parameter for Age_gt30 dummy variable: the estimate is positive and statistically significant at the 0.01 level. This suggests that users who are older than 30 tend to study slightly longer than the younger users.
- The parameter for the interaction variable between treatment dummy and age_gt30 dummy: it is negative and statistically significant. The result suggests that the treatment effect is slightly lower for the older users.

Summary

To evaluate the causal effect of an intervention, the core idea is to create a comparison that mimics a parallel universe, where everything else is the same except for the presence or absence of the intervention. The most widely used method to approximate this ideal comparison is the randomized controlled trial (RCT).

In an RCT, participants are randomly assigned to either a treatment group (which receives the intervention) or a control group (which does not). Before running the experiment, it is important to confirm that the two groups are statistically similar in their baseline characteristics. This is often done by performing t-tests on observed variables to check for any pre-existing differences.

Once the experiment is completed and outcome data are collected, two common types of analysis are typically conducted:

- Average Treatment Effect (ATE): Estimated using a t-test to compare mean outcomes between the treatment and control groups.
- Heterogeneous Treatment Effects (HTE): Estimated using multivariate regression models to explore how the treatment effect varies across subgroups defined by user characteristics.

Homework

1. Suppose a company wants to test whether sending daily motivational emails increase user engagement. Identify two key conditions that must be satisfied for the random assignment to be valid.
2. You run a two-sample t-test comparing weekly study hours between treatment and control groups and obtain a *p*-value of 0.27.
   (a) State the null and alternative hypotheses.
   (b) Based on the *p*-value, what conclusion should you draw?
   (c) Would this result support or undermine the validity of the RCT? Explain.
3. Consider the regression model in the chapter that includes: a treatment dummy (Treat_dm), an age-over-30 dummy (Age_gt30), and an interaction term between them.
   (a) What does the coefficient on the interaction term measure?
   (b) Interpret what it means if this interaction coefficient is negative and statistically significant.

CHAPTER 12

Artificial Intelligence in Marketing Analytics

Learning Objectives

1. Understand the evolution of artificial intelligence (AI).
2. Identify key applications of large language models in marketing analytics.
3. Apply the Gemini API to analyze online customer reviews.

What Is Artificial Intelligence?

Artificial intelligence (AI) refers to the development of computer systems capable of performing tasks traditionally associated with human intelligence. These tasks include learning, reasoning, problem-solving, and many more. The list of tasks can be dedicated to an AI system or agent is growing everyday, as the technology progresses. In recent years, AI has become a transformative force across industries, enabling more personalized, data-driven, and scalable decision-making. In marketing, AI powers recommendation systems, customer segmentation, and, increasingly, the generation and analysis of unstructured data such as text, images, and videos.

To fully appreciate the impact of AI in marketing, it is essential to begin with a historical understanding of how the field of artificial intelligence emerged and evolved.

Supplementary Information The online version contains supplementary material available at https://doi.org/10.1007/978-3-032-11130-2_12.

© The Author(s), under exclusive license to Springer Nature Switzerland AG 2026

X. Dong, *Marketing Analytics and Data Science*,
https://doi.org/10.1007/978-3-032-11130-2_12

The Birth of AI: A Historical Overview

The aspiration to build machines that emulate human thought dates back centuries. However, the term "artificial intelligence" was formally coined in the summer of 1956 at a seminal workshop held at Dartmouth College. Known as the Dartmouth Summer Research Project on Artificial Intelligence, the event was organized by John McCarthy (Stanford, IBM), Marvin Minsky (MIT), Nathaniel Rochester (IBM), and Claude Shannon (MIT). This workshop is widely regarded as the founding moment of AI as an academic discipline, with its organizers recognized as the field's founding figures.

Yet the intellectual roots of AI go deeper. One of the most influential early thinkers was Alan Turing, a British mathematician and logician, whose work in the 1930s–1950s laid the foundation for modern computing and AI.

Turing is often considered a foundational figure in computer science. His story, including his contributions during World War II in breaking the German Enigma code, was popularized in the film *The Imitation Game* (2014). More profoundly, Turing's legacy lies in his theoretical exploration of whether machines can exhibit intelligence.

In 1950, Turing published the landmark paper "Computing Machinery and Intelligence" in the journal *Mind*. He famously posed the question: "Can machines think?" To explore this question, he introduced a thought experiment now known as the Turing Test.

The Turing Test

Rather than wrestling with the philosophical ambiguity of whether a machine can "think", Turing proposed a more practical question: Can a machine imitate human responses so well that an interrogator cannot reliably distinguish it from a human?

In the Turing Test, a human judge conducts a conversation via text with both a human and a machine. If the judge cannot reliably identify which participant is the machine, the machine is said to have passed the test. Although the Turing Test has been critiqued over time, it remains a symbolic benchmark in AI, highlighting key issues in language understanding, reasoning, and learning.

The Historical Evolution of AI: Four Key Waves

Since the 1950s, AI has undergone cycles of innovation and stagnation, often referred to as "AI winters" and "AI booms". These waves reflect technological advancements and shifts in both academic and commercial enthusiasm.

1. Rule-Based Systems and the First AI Winter (1950s–1970s)
   Early AI systems relied on manually crafted rules. Although they worked in limited domains, they were unable to adapt to the complexity and ambiguity of real-world problems. By the 1970s, performance limitations and overhyped expectations led funding agencies (e.g., DARPA and the UK government) to scale back support, triggering the first AI winter.
2. Expert Systems and the Second AI Winter (1980s)
   AI experienced a resurgence in the 1980s with the emergence of expert systems, which consist of programs that mimic decision-making using large sets of hand-coded rules. These systems were widely adopted in business but eventually became unwieldy as rule sets grew. Their lack of scalability led to another decline in interest by the late 1980s, known as the second AI winter.
3. Rise of Machine Learning (1990s–2000s)
   In the 1990s, AI shifted from rule-based systems to data-driven approaches. Machine learning has emerged as the dominant paradigm, enabling computers to recognize patterns by training models on large datasets. Improvements in data availability and computing power made this possible. In 1997, IBM's Deep Blue famously defeated chess champion Garry Kasparov, demonstrating the growing capabilities of AI.
4. Deep Learning and Large-Scale Models (2010s–Present)
   The 2010s saw a revolution in deep learning, a subfield of machine learning that uses multi-layer neural networks to model complex patterns. Breakthroughs in natural language processing, computer vision, and speech recognition were fueled by advances in GPU hardware and access to large datasets.

Notably, the release of transformer-based models such as GPT-3 (2020) by OpenAI and Gemini (2023) by Google dramatically expanded AI's capabilities. These large language models (LLMs) can generate human-like text, translate languages, and answer questions, tasks once considered beyond the reach of machines. Compared with earlier generations of AI, today's large language models (LLMs) can not only interpret and analyze human input in many formats, but also generate content in many formats, such as text, images, audio, and even video. Importantly, they can also produce computer-executable programming code. This ability allows LLMs to instruct computers to perform a wide range of tasks, including ones that previously required substantial and carefully engineered software to automate. As these capabilities continue to expand, they have sparked public debate and concern about whether AI might replace human workers. Personally, I believe the more productive approach is to understand these tools and use them in a way to improve efficiency and decision making. A broader discussion of AI's societal implications is important, but it is beyond the scope of this book.

AI in Marketing Analytics

Throughout this book, we introduce a range of data-driven approaches for solving marketing problems. Many of these rely on structured numerical data, such as click rates, sales numbers, ad budgets, and user demographics. These are typically analyzed using traditional statistical and machine learning models.

However, many marketing activities involve unstructured data, such as text from customer reviews, images from product listings, or video content in advertisements. Historically, analyzing such content was extremely challenging due to the difficulty of converting it into a numerical format.

Recent advances in LLMs and image/video understanding now allow marketers to extract meaning from this unstructured content. For example, a marketing manager at Nike can now ask:

- How do customers describe the comfort of our new running shoes in reviews?
- Do images in our social media ads align with our brand message?
- Which product descriptions drive the most engagement?

To answer these questions, marketers can now use LLMs not only as analytical tools but also as creative collaborators.

What are LLMs

The seminal paper "Attention is all you need" published in 2017 (Vaswani et al. [4]) introduced the Transformer architecture that became a major breakthrough in modern language modeling. In our daily conversations, understanding human language requires tracking context. The same word may have different meanings depending on its surrounding words and sometimes distant references earlier in a sentence or document. Previous language models such as recurrent neural networks (RNNs) struggled to capture these dependencies, especially when it is long-range. The Transformer approach addressed this limitation by developing the self-attention method. Self-attention allows the model to assign different weights to different words (represented as tokens) when forming a representation of each word or token, so that it can directly connect related words even when they are far apart in the input. This design both improves the modeling of long-distance relationships and allows highly parallel computation during training.

Leveraging the transformer architecture, researchers and technology companies have developed large language models (LLMs) capable of processing and generating human-like text, interpreting images, and even understanding video content. These models are trained on massive datasets that include text, images, and video collected from across the Internet. Since machines cannot directly interpret raw language or visual data, the transformer framework first encodes this unstructured information into numerical form, representing words, pixels, or audio as vectors. The model then learns patterns and relationships among these encoded vectors, enabling it to generate meaningful and contextually appropriate responses.

The scale of these models is genuinely unprecedented. Most of the models introduced earlier in this book, such as logistic regression, typically involve only a few dozen parameters. In contrast, GPT-3, one of the most well-known LLMs developed by OpenAI, was trained with approximately 175 billion parameters [2]. These parameters capture the associations between sophisticated linguistic patterns, semantic structures, and contextual clues at an extraordinary level of sophistication.

As LLMs can understand human instructions and generate both natural language and computer-executable code, they can be used to automate many tasks, such as searching the web, summarizing information, and instruct the computer to do more tasks using AI. In other words, when we provide clear, well-specified prompts, LLMs can help complete tasks that is time-consuming for us. This prompts the rise of Agentic AI, where AI can serve as our agents by itself or as an organized team to perform many tasks on our behalf. It is therefore critical to use LLM-based AI systems responsibly. As these technologies advance rapidly, clear governance through policies, standards, and regulations will be important to define appropriate boundaries for their use and to reduce potential harms.

While marketing practitioners don't need to understand every technical detail of how LLMs work, unlike simpler models such as linear regression, it is essential to recognize their enormous capbilities. LLMs can now perform tasks that were once infeasible, such as summarizing long-form text, extracting sentiment and product attributes from customer reviews, interpreting the content of images, and generating creative designs. LLM acts less like a statistical model in traditional machine learning, instead, it is more like a general-purpose infrastructure, similar to how we use the Internet or cloud computing.

Training such a model, however, is a highly resource-intensive process. It involves:

- Gathering vast quantities of data, including both publicly available online content and licensed or copyrighted sources like books and academic journal articles.
- Assembling the necessary hardware infrastructure to process and learn from this large amount of data. This requires significant computing power, typically through the use of GPU (graphics processing unit) clusters. Nvidia's V100 GPU, for example, has been widely used for training LLMs. According to Narayanan et al. [3], training a model the size of GPT-3 would take approximately 288 years on a single V100 GPU, making it impractical for individual or small companies to build.

Due to the formidable cost and complexity of developing LLMs, only a few organizations can build them. Fortunately, once these models are trained, they can be made accessible to others via cloud-based APIs (Application Programming

Interface). This means businesses and researchers can leverage the power of LLMs for specialized tasks without needing to build or train the models themselves.

Accessing LLMs for Marketing Applications

As of early 2026, several LLMs are publicly available. Some are open-source, while others are proprietary and require payment for usage. Pricing typically depends on two factors: the number of tokens you input into the model (i.e., the amount of information the model must process) and the number of tokens it generates in response (i.e., the length and richness of the output).

Here is a list of the popular models (which could be changing over time):

- GPT (OpenAI): https://openai.com
- Claude (Anthropic): https://www.anthropic.com
- Gemini (Google DeepMind): https://deepmind.google
- Grok (xAI): https://x.ai/
- LLaMA (Meta): https://www.llama.com

As advancements in hardware and software continue to accelerate, large language models (LLMs) are becoming increasingly powerful, affordable, and accessible. These tools are improving every day, and the range of tasks they can handle is expanding rapidly. In this section, we use a simple example to demonstrate the most basic way to connect to an AI tool and begin leveraging its capabilities. Although the example is intentionally simple, it aims to serve as a practical starting point. As these systems continue to evolve, the ability to connect to AI tools will remain a foundational skill.

One use case involves customer reviews collected from e-commerce platforms. These reviews are typically unstructured text, making them difficult to incorporate into conventional machine learning models. However, with LLMs, we can input thousands of such reviews and ask the model to extract structured insights, such as product attributes, sentiment, or user preferences, and return the results in a tabular format suitable for further analysis.

To perform this kind of task using an LLM, two types of resources are required:

1. Access to a Large Language Model
   First, you need to obtain access to an LLM. These models are accessed through cloud platforms or API services provided by companies such as OpenAI, Google, or Hugging Face. You interact with the model by sending a prompt that includes both the data (e.g., a set of reviews) and instructions (e.g., extract product sentiment or highlight key features).

LLMs process inputs in the form of tokens, which are units of text such as whole words or word fragments. You are typically charged based on the number of input tokens (i.e., the amount of information the model needs to read) and output tokens (i.e., the amount of text it generates in response to your prompts). To connect your program to an LLM, you will need to use an Application Programming Interface (API), which provides a standard way for your code to communicate with the model over the Internet.

2. Computational and Deployment Resources

   Second, LLMs require substantial computational power to operate, particularly for tasks that involve large datasets or lengthy responses. Fortunately, this need is abstracted away by cloud providers such as Amazon Web Services (AWS), Microsoft Azure, and Google Cloud. These platforms host the models on powerful hardware and offer flexible pricing based on usage. As a user, you simply send requests to the hosted model, and the cloud provider manages the processing and infrastructure behind the scenes.

Using Gemini to Label Customer Reviews: An Example

Suppose you have a dataset of customer reviews from Amazon. Traditionally, analyzing thousands of text entries would be difficult using standard machine learning models. However, with an LLM, you can process each review and extract structured insights from a large amount of unstructured texts, such as how customers rate a product's comfort, weight, and appearance, from each review.

For demonstration purposes in this textbook, we use Gemini 2.0 Flash in the example, as it offers free access to both LLM deployment and cloud services, such as GPUs and storage, under certain limits.

Step 1: Set Up Free Access with Google AI Studio

- Visit https://aistudio.google.com and sign in with your Google account.
- Create an API key by clicking the button at the top-right corner of the page after you log in. This is a personal credential that allows your code to access Gemini's services and enables Google to track the resources you have been using.
- Use your API key in a script to analyze review content.

Step 2: Analyze the Product Review

In the following example code, we demonstrate how to use Gemini to analyze a customer review from Amazon for a pair of sneakers. The goal is to extract structured insights across three dimensions: comfort level, weight, and appearance.

- The text of the product review is stored in a variable called review.
- The api_url specifies the target endpoint, including the model name and your API key.
- The prompt variable defines what you want Gemini to do, in this case, categorize the review according to the three product attributes.
- The body_data object defines the input structure required by Gemini's API. This follows a nested JSON format, where each interaction includes one or more "parts". In this example, we use a single part consisting of the prompt text.

Once this input is assembled, we send an HTTP POST request to the Gemini API using the POST command. Here's how the main components of the request work:

- url: it is the endpoint that includes your API key and specifies which Gemini model to use.
- add_headers: it tells the server that we are sending data in JSON format.
- body: it converts the structured R list (body_data) into a valid JSON string using jsonlite::toJSON().

Together, these elements prepare and execute a properly authenticated request to Gemini's inference engine, enabling you to extract structured insights from unstructured text programmatically.

```
library(httr)

## Warning: package 'httr' was built under R version 4.4.3

library(jsonlite)

## Warning: package 'jsonlite' was built under R version 4.4.3

library(dplyr)

##
## Attaching package: 'dplyr'

## The following objects are masked from 'package:stats':
##
##     filter, lag

## The following objects are masked from 'package:base':
##
##     intersect, setdiff, setequal, union

library(readr)

## Warning: package 'readr' was built under R version 4.4.3

api_url <- "https://generativelanguage.googleapis.com/v1beta/models/gemini-2.
0-flash:generateContent?key=AIzaSyDzHyGAb0ARq_ij6x6DvuwxTN499uOIaGc"

# Sample product review
review <-"
I was drawn to these ALLSWIFIT sneakers because they're super cute and look v
ery similar to some of the more expensive brands. I ordered them in two color
s because I loved the style so much! The pink/orange combo is really vibrant
and fun. My initial impression is that they're very lightweight and breathabl
e, which is a plus. However, I did notice that they're a little bit loose on
my feet. I'm a true size 10, and while these fit length-wise, there's some ex
tra room. I'm hesitant to size down, though, because I worry that a smaller s
ize might be too tight around the ankle and footbed, especially since some ot
her reviewers mentioned the opening being snug. I'll probably add some non-sl
ip heel pads to see if that helps with the looseness. Despite the slight sizi
ng issue, I'm still happy with them so far. They're comfortable enough for ca
sual wear, and I'm hoping the heel pads will solve the minor fit problem. For
 the price, they're a great dupe for the more expensive styles.
"


# the prompt as input to the LLM
prompt <- paste0(
    "Please analyze the following product review and provide labels for:\n",
    "1. Comfort Level (High/Medium/Low)\n",
    "2. Weight of the Shoes (Light/Medium/Heavy)\n",
    "3. Appearance (e.g., Stylish/Vibrant/Plain/Dull)\n\n",
    "Review:\n", review #, "\n\nLabels:"
)
```

```
# build a nested structure to meet Gemini's input format. It will be converte
d to a JSON object and sent to the API.
body_data <- list(
    contents = list(
      list(
        parts = list(
          list(text = prompt)
        )
      )
    )
  )

# It sends an HTTP POST request to the Gemini API
res <- POST(
  url = api_url,
  add_headers(
    `Content-Type` = "application/json"
  ),
  body = toJSON(body_data, auto_unbox = TRUE)
)
```

Now, after calling Gemini to analyze the user review, the output generated by Gemini can be printed out, as below. Among them,

- status_code(response) checks the numeric HTTP code (e.g., 404). If the returned code is 200, it indicates that the task was successful. Otherwise, it goes to else and prints out the error. If successful, the contents of the response will be written to the variable result by parsing the JSON response into an R list.
- result$candidates[[1]]$content$parts[[1]]$text contains the actual string the model generated.

```
# print out the raw output from Gemini

if (status_code(res) == 200) {
  result <- content(res, as = "parsed", type = "application/json")
  # the output from the Gemini model
  output_text <- result$candidates[[1]]$content$parts[[1]]$text
  cat("Raw Output:\n", output_text, "\n")

} else {
  cat("Request failed with status code", status_code(res), ":\n", content(res
, as = "text"))
}

## Raw Output:
##  Here's the analysis of the product review based on your requested labels:
##
## 1.  **Comfort Level:** Medium
##
## *   *Reasoning:* The reviewer states the shoes are "comfortable enough for
 casual wear," which suggests a moderate level of comfort rather than excepti
onal comfort (High) or discomfort (Low). The sizing issue also affects the co
mfort level.
##
## 2.  **Weight of the Shoes:** Light
##
## *   *Reasoning:* The reviewer explicitly mentions "they're very lightweigh
t."
##
## 3.  **Appearance:** Stylish/Vibrant
##
## *   *Reasoning:* The reviewer states they are "super cute" and "loved the
style so much!" Also, the "pink/orange combo is really vibrant and fun."
##
```

The raw output returned by Gemini is not in a structured format; it appears as plain text with labels such as Comfort Level, Weight of the Shoes, and Appearance, each followed by a descriptive value (e.g., "Medium"). To organize this output into a structured format such as a data frame, we need to extract only the values associated with each label, using the following R script.

```
library(stringr)

## Warning: package 'stringr' was built under R version 4.4.3

    # Try to parse the labels
    comfort <- str_match(output_text, "\\*\\*Comfort Level:[\\*\\-\\(\\s*]*\\
**([A-Za-z/]+)\\**")[,2]
    weight  <- str_match(output_text, "\\*\\*Weight of the Shoes:[\\*\\-\\(\\
s*]*\\**([A-Za-z/]+)\\**")[,2]
    appearance <- str_match(output_text, "\\*\\*Appearance:[\\*\\-\\(\\s*]*\\
**([A-Za-z/]+)\\**")[,2]

    comfort

## [1] "Medium"

    weight

## [1] "Light"

    appearance

## [1] "Stylish/Vibrant"
```

To achieve this, we use regular expressions in R with the str_match() function from the stringr package. This function enables us to search for patterns in text and extract specific substrings that match the defined pattern.

The general syntax is str_match(string,pattern)[,2], which extracts the first captured group that matches the pattern inside the string.

Let's look specifically at the regular expression used to extract the comfort level:

```
"\\*\\*Comfort Level:[\\*\\-\\(\\s*]*\\**([A-Za-z/]+)\\**"

## [1] "\\*\\*Comfort Level:[\\*\\-\\(\\s*]*\\**([A-Za-z/]+)\\**"
```

In the above pattern:

- ** indicates the pattern with two asterisks next to each other **.
- Comfort Level specifies the exact words we are looking for in the output string.
- [*\\-\\(\\s*]* matches any one of them inside the bracket, including *, - or any white space. It is necessary to have these options, as the model outputs sometimes have slightly different formatting. The last asterisk allows matching one or more of those included in the bracket. - **([A-Za-z/]+)** matches any letters enclosed by the asterisks.

Altogether, this pattern looks for a string like **Comfort Level:** Medium** and extracts only the word "Medium", excluding the label and formatting symbols.

The final use of [,2] selects the component returned by str_match(), which contains the text captured by pattern in the string.

This approach can be repeated for other fields, such as Weight and Appearance, by changing the label in the pattern (e.g., replacing "Comfort Level" with "Weight of the Shoes" or "Appearance").

By extracting just the values for each label, we can then assemble them into a data frame and prepare the data for further analysis or visualization. To put them into a formatted data.frame, we can use the following R script.

```
label_df <- data.frame(
  review_text = review,
  Comfort_Level = comfort,
  Weight = weight,
  Appearance = appearance,
  stringsAsFactors = FALSE
)

# Print the data frame
print(label_df)
```

```
## review_text
## 1 \nI was drawn to these ALLSWIFIT sneakers because they're super cute and
 look very similar to some of the more expensive brands. I ordered them in tw
o colors because I loved the style so much! The pink/orange combo is really v
ibrant and fun. My initial impression is that they're very lightweight and br
eathable, which is a plus. However, I did notice that they're a little bit lo
ose on my feet. I'm a true size 10, and while these fit length-wise, there's
some extra room. I'm hesitant to size down, though, because I worry that a sm
aller size might be too tight around the ankle and footbed, especially since
some other reviewers mentioned the opening being snug. I'll probably add some
 non-slip heel pads to see if that helps with the looseness. Despite the slig
ht sizing issue, I'm still happy with them so far. They're comfortable enough
 for casual wear, and I'm hoping the heel pads will solve the minor fit probl
em. For the price, they're a great dupe for the more expensive styles.\n
##   Comfort_Level Weight       Appearance
## 1        Medium  Light Stylish/Vibrant
```

Suppose you have a dataset containing a large collection of user reviews. In that case, you can apply the code described above iteratively by processing each review individually and appending the extracted results as a new row in a dataframe. This allows you to transform a collection of unstructured text reviews into a structured, tabular format containing clearly defined labels.

The specific labels you extract can be customized to suit your analytical goals. For instance, instead of extracting attributes like comfort level, weight, and appearance, you could modify the prompt to identify other product features such as color, durability, sizing accuracy, or customer sentiment. The flexibility of large language models enables you to tailor the analysis to align with your particular research or business objectives.

This simple example illustrates how to connect with Gemini API, the same workflow applies to other LLM APIs as well. In fact, even the code to implement the task can be generated with a LLM. However, as these tools become more capable, it remains essential to understand the underlying fundamentals of how they work.

Summary

End of the Book

Congratulations on reaching the end of the book! As this book is about data-driven marketing decisions, it closes by highlighting one of the most recent technologydevelopment, i.e. Artificial Intelligence, and its application in in conducting Marketing Analytics. Marketing is a dynamic field. As new technologies emerge, they reshape how consumers discover, evaluate, and interact with businesses. They also require firms to adapt and to develop new ways to undestand customers in order to make better marketing and business decisions. This book is meant to be a starting point for that journey, equipping readers with foundational concepts and practical tools they can build on as the technology and marketing continue to evolve.

Earlier in this book, we introduced a range of data analytics tools commonly used to solve marketing problems. However, those tools primarily focus on structured data, including numerical or categorical variables organized in tabular format. In real-world marketing, much of the valuable information exists in unstructured formats, such as customer reviews, social media posts, images, and videos. These sources often reflect user sentiment, brand perception, and responses to marketing efforts, but they have historically been challenging to analyze using traditional models.

The emergence of large language models (LLMs) has changed that. LLMs have revolutionized many fields by demonstrating their ability to understand and generate human language, enabling tasks once considered out of reach for automation. Although training these models requires enormous computational resources and is currently feasible only for a small collection of institutions, their availability through APIs has made them accessible to a much broader audience. For marketing professionals, pre-trained LLMs can now serve as an infrastructure upon which specialized applications can be built.

One such application is the ability to extract structured information from large volumes of unstructured customer review data. Rather than manually coding or discarding this information, we can now use LLMs to convert free-text reviews into structured, analyzable formats.

To make this process even more accessible, this chapter demonstrated how to use Google AI Studio, a platform that offers free access to a Gemini model. With only a Google account and an API key, you can deploy an LLM in the cloud and process review data without any cost or local installation. The example R code provided in this chapter and the Python code from the website show how to submit a review to the model, extract relevant attributes (such as comfort, weight, and appearance), and organize the results into a data frame. From there, the output can be analyzed using the statistical and machine learning tools discussed earlier in the book.

Although current free-tier access may be limited in scale, continued advancements in both AI models and cloud infrastructure will likely make LLMs even more affordable and accessible in the near future. This opens up exciting opportunities for marketers to harness unstructured data at scale, turning text, images, and videos into actionable insights [1].

Homework

1. Follow the instructions and create an API key from Google AI Studio.
2. Find a product you are interested in from Amazon, copy a long customer review into your code, use Gemini to extract information from the customer review, and put the output into a data format.

Reference

1. Turing, A. M. (1950). Computing Machinery and Intelligence. *Mind, 59*(236), 433–460. https://doi.org/10.1093/mind/LIX.236.433
2. Brown, T., Mann, B., Ryder, N., Subbiah, M., et al. (2020). Language Models are Few-Shot Learners. arXiv preprint arXiv, 2005.14165, 2020.
3. Narayanan, D., Shoeybi, M., Casper, J., et al. (2021). Efficient Large-Scale Language Model Training on GPU Clusters Using Megatron-LM. arXiv, 2104.04473.
4. Vaswani, A., Shazeer, N., Parmar, N., Uszkoreit, J., Jones, L., Gomez, A. N., Kaiser, L., & Polosukhin, I. (2017). Attention Is All You Need. 31st Conference on Neural Information Processing Systems (NeurIPS). Long Beach, CA.

The manufacturer's authorised representative in the EU is Springer Nature Customer Service Centre GmbH, Europaplatz 3, 69115 Heidelberg, Germany. If you have any concerns regarding our products, please contact ProductSafety@springernature.com

Printed and bound by CPI Group (UK) Ltd, Croydon, CR0 4YY
22/09/2026
02220981-0005